Messerschmitt Me 109

in Swiss Air Force Service

Georg Hoch

Schiffer Military History
Atglen, PA

Book translation by Christine Wisowaty

Book Design by Ian Robertson.

Copyright © 2008 by Georg Hoch.
Library of Congress Control Number: 2008922671

All rights reserved. No part of this work may be reproduced or used in any forms or by any means – graphic, electronic or mechanical, including photocopying or information storage and retrieval systems – without written permission from the copyright holder.

Printed in China.
ISBN: 978-0-7643-2924-1

This book was originally published in German under the title
Die Messerschmitt Me 109 in der Schweizer Flugwaffe - ein Stück Zeitgeschitchte

Published by Schiffer Publishing Ltd.
4880 Lower Valley Road
Atglen, PA 19310
Phone: (610) 593-1777
FAX: (610) 593-2002
E-mail: Info@schifferbooks.com.
Visit our web site at: www.schifferbooks.com
Please write for a free catalog.
This book may be purchased from the publisher.
Please include $5.00 for shipping.
Try your bookstore first.

In Europe, Schiffer books are distributed by:
Bushwood Books
6 Marksbury Avenue
Kew Gardens
Surrey TW9 4JF, England
Phone: 44 (0) 20 8392-8585
FAX: 44 (0) 20 8392-9876
E-mail: Info@bushwoodbooks.co.uk.
Visit our website at: www.bushwoodbooks.co.uk
Free postage in the UK. Europe: air mail at cost.
Try your bookstore first.

Contents

	The Quotation
	The Author's Thanks
	Foreword
Chapter 1	The Development of the Fighter Aircraft in the Thirties 11
Chapter 2	The Aircraft Messerschmitt Me 109
	Prehistory 12
	The Me 109 – The Multipurpose Aircraft 13
Chapter 3	Willy Messerschmitt the Person 14
	Messerschmitt BFW 109, Bf 109 or Me 109? 15
Chapter 4	The Me 109 at the International Aviation Meeting in Dübendorf 1937 ....
	International Debut of the Me 109 16
	No Real Competition for the Messerschmitt 109 18
	The Cause of the Accident of the Me 109 V14/D-ISLU 24
Chapter 5	The Procurement of Aircraft in the Thirties
	The Long Path to New Aircraft 26
	The He 112 with the Hispano Engine 29
	Dark Clouds above Europe and Still No New Fighter Aircraft 31
Chapter 6	The Procurement of the Me 109 D/E
	Hermann Göring and the Stag Hunting 34
	The Controversial Procurement of the Second Series 35
	The Delivery of the Me 109 – The Bone of Contention between Switzerland and the German Reich 37
Chapter 7	The Weapons of the Me 109 D and E
	Ambiguous Perceptions of the Armament 39
	The Weapons of the Me 109 D 40
	The Weapons of the Me 109 E 43
	The Cockpit of the Me 109 E 48
Chapter 8	The Radio Equipment of the Me 109 D and E
	Unclear Responsibilities 51
	Questionable Equipment Procurement 51
	The Problem with the S.F.R. Equipment 53
Chapter 9	Introduction and Operation of the Me 109 D and E
	The New Aircraft Me 109 D – New Problems 56
	Me 109 E A New Challenge 61
	Messerschmitt – Lightweight Construction with Defects 65
	Increased Efficiency with the Escher-Wyss Propeller 72
	The Signal Rocket Equipment 73
	War Gasoline (Kriegsbenzin) 73
	The Instillation of the Explosion-proof Oxygen System 76
	The Emil becomes a Jagdbomber 77
	Antitank Protection in the Me 109 E 78
	The Night Flight Equipment 79
	Age and Decrepitude, Swimmer and Arrestor Hooks 80
	The Me 109 E from Spare Parts 81

Chapter 10	The Aircraft Procurement in the Forties
	The Search for a Modern Fighter Aircraft 83
	Controversial Licensed Manufacture for the German Reich 83
	The Morane is Classified as Unfit for War 84
	The Macchi C 202 84
	Negotiations with Germany 85
	Divisionär Bandi is Discharged 86
	No Chance for Proprietary Development 86
Chapter 11	The Events from September 1939 until June 1940
	The Me 109 A Modern Aircraft at the Right Time 89
	Deceptive Peace 91
	The First Shots on Intruders 95
	Problems with the Border in Jura 98
	June 2, 1940 – The Odyssey of Uffz. Mahnert ..101
	June 4, Göring's Revenge 104
	June 8 – The Boundary is Overstepped 108
	Outline Map Events in May – June 1940 114
Chapter 12	The Events as of Summer 1940
	Flyers with Propped Wings 115
	The Reintroduction of the Jagdpatrouille 116
	The Avenches Base in Summer 1940 118
	The Everyday Life of the Messerschmitt Units .124
	The 5th of September and its Consequences 132
	The Decommission of the Me 109 D and E 142
	Outline Map "Events as of Summer 1940" 143
Chapter 13	The Me 109 F ....
	The Emergency Landing in Bern-Belp 144
	The Internment of Martin Villing and Heinz Scharf 145
	The Me 109 in the Swiss Flugwaffe 145
Chapter 14	The Procurement of the Me 109 G-6
	1944 and Still No Modern Aircraft 149
	Interlude in Samedan 149
	The Mysterious Nachtjäger 150
	A Special Topic: The Production of the Me 109 G 151
	The Me 109 G-6 for the Swiss Flugwaffe 153
	Munitions Procurement 154
	Antennas and Radio Equipment 154
	Erla Hood and Augmented Rudder Assembly 155
	Additional Tanks and Bombs 156
	Unforeseen Gain 156
	One Last Interest for the Me 109: the Me 109 G-10/AS 158
	Me 109 G – Engines and Manufacturers 160
Chapter 15	The Me 109 G in Action
	The Disappointed General 161
	Unknown Sounds 161
	The Engine Problems 162
	Fluttering Stabilizers 162
	The Forgotten J-713 170
	The End of the Me 109 G 171
	The Me 109 G – Better than its Reputation 174

Chapter 16	Camouflage and Markings	
	The Camouflage...................................175	
	National Emblem and Markings.......179	
	Me 109 F – Open Question 181	
	Kompanie Emblem............................183	
	What was published next...! 190	
Chapter 17	The Whereabouts of the Me 109 in Switzerland	
	The Pilatus P-2..................................193	
	The J-355 in the Dübendorf Fliegermuseum....193	
	Assemblies, Weapons and Engines 194	
	The Remnants of the J-310 195	
Chapter 18	*Color Photographs* 196	
	Color Profile.....................................200	
	Aircraft Markings, National Emblems and Labeling..................................207	
Appendix 1	Me 109 D Weapon Installation MG 17 212
Appendix 2	Me 109 E Cowlings of DB 601 and MG 29.....213
Appendix 3	Me 109 E MG 29 – Weapon Installation 214
Appendix 4	Me 109 E MG 29 – Ammunition Box / Casing Box...215
Appendix 5	Me 109 E Control Unit for MG 29 216
Appendix 6	Me 109 E Cable Pulley for MG 29 217
Appendix 7	Me 109 E Assembly FF-K 218
Appendix 8	Me 109 E Electrical Equipment for FF-K 219
Appendix 9	Me 109 E Cable Pulleys for FF-K 220
Appendix 10	Me 109 E Control Stick KG 11 221
Appendix 11	Me 109 E Control Stick Contact Pushbutton ...222
Appendix 12	Me 109 E Aiming Device "Revi 3c" 223
Appendix 13	Me 109 E Signal Rocket Equipment 224
Appendix 14	Me 109 E Bomb Equipment.............................225
Appendix 15	MG 29 Adjustment Table for Synchronization 226
Appendix 16	Me 109 E Comparison VDM – EQ V6 Propeller 227
Appendix 17	Bull's Eye Target He 111 1G+HT, Kemleten May 16, 1940..228
Appendix 18	Bull's Eye Target He 111 G1+HS, Ursins June 2, 1940...229
Appendix 19	Avenches Airport, Summer 1940....................230
Appendix 20	Interlaken Airport, April 1942 231
Appendix 21	Zweisimmen Airport, December 1942 232
Appendix 22	Unterbach Airport, July 1943 233

Messerschmitt Me 109 Units from 1939-1949...234
Messerschmitt Me 109 Flight Statistics 235
Messerschmitt Me 109 Dimensions and Performance Characteristics.....................236
Messerschmitt Me 109 Dates 237
Abbreviations 252
Index...253
Sources/Bibliography 256

The Quotation

"Full of wonder, I saw what Swiss spirit and circumspection had created here. May Dübendorf always remain a stronghold from which the bold Swiss flyers rise to the glory and honor of their fatherland!

In gratitude and camaraderie: Hermann Göring, *Hauptmann* a. D., lastly *Kommandeur* of the *Jagdgeschwader* 'Freiherr v. Richthofen' No. 1 during the World War."

Göring's entry in the guest book of the officers' mess in Dübendorf, November 1927.
In June 1940 he commanded his pilots against the Swiss—the audacious Swiss flyers hit back.

The Author's Thanks

To write a study of the Me 109 in the Swiss *Flugwaffe* was only possible thanks to the help of Federal Offices, private individuals, and former and active members of the air force. Special thanks goes to Mr. Toni Bernhard, conservator of the *Fliegermuseum* in Dübendorf until Spring 1998. Special thanks also goes to Fa Dietschi, AG printing and publishing house in Waldenburg, who made it possible to realize my ideas and produce my book at a reasonable price. Thanks to my wife, Franzisca, who had the necessary understanding for the preceding time.

Whether on the search for documents, footage, memories, or review of the manuscript, numerous competent people assisted:

G. Braun	B.F.H. Bavarian Aircraft Historian e.V.	H. Ruegge	ex Fl Kp 15/Instructor
W. Bürli	ex DMP	Frau Rufer	Widow of Hptm. A. Rufer Fl Kp 15 and Fl Kp 9, Pilot
A. Cueni	ex DMP	P. Schmoll	
F. Dannecker	ex Fl Kp 8 Pilot	H. Siegfried	ex Fl Kp 7 Pilot
K. Ernst		H.H. Stapfer	
H.P. Häberlin	ex Fl Kp 9 Pilot	K. Sturzenegger	ex Fl Kp 21/UeG Pilot
K. Hänggi		M. Villing	ex JG 5 Pilot, Germany
C. Heer	ex DMP	A. Violand	
R. Heiniger	ex Fl Kp 7 Pilot	M. Weichelt	Landesbibliothek Bern
F. Heller	ex DMP	F. Wiesendanger	ex Fl Kp 7 Pilot
Ph. Hilt	B.F.H. Bavarian Aircraft Historia e.V.	E. Zeller	ex Doflug Altenrhein
H.R. Hofer	ex Fl Kp 7 Tech Of		
R. Homberger	ex Fl Kp 15 Pilot	Armee-Bibliothek	Bern
M. Kägi	ex DMP	Armee-Foto-Dienst	Bern
R. Küng		Bundesarchiv	Bern
E. Künzler		Fliegermuseum	Dübendorf
A. Lareida	Conservator, Dübendorf Flieger Museum	Fotodienst	BABLW Dübendorf
H. Lohman	ex *Einflieger* for Messerschmitt	Fotodienst	SF Buochs
R. Muggli		Fotodienst	SW Thun
A. Muser	ex Fl St 21/Swissair Pilot	Service de foto	OFEFA Payerne
R. Renfer			

Numerous photographs originate from private sources. The authors are partly known. Thanks to everyone who provided me with pictures from their albums, collections, or single pictures for the illustration of this book: G. Braun, Ph. Clot, A. Cueni, F. Dannecker, H.R. Dubler, K. Ernst, St. Gfeller, Ch. Gloor, H.P. Häberlin, K. Hänggi, A. Hayoz, R. Heiniger, F. Heller, Ph. Hilt, R. Homberger, A. Kehrli, R. Küng, R. Muggli, F. Rapin, H. Ruegge, Frau Rufer, P. Schmoll, H.H. Stapfer, K. Sturzenegger, M. Villing, A. Violand, M. Weichelt, and E. Zeller.

Foreword

In the history of aviation, no other aircraft has struck such a wide interest as the Messerschmitt Me 109. Those parties less interested in aviation can imagine the Messerschmitts to be of significance. At the International Aviation Meeting in Dübendorf in 1937, during the Civil War in Spain, or in the campaign against Poland and France, the superiority of this aircraft was demonstrated. German propaganda added to this. In the air battles for Dunkirk at the end of May 1940, and in the battle for England, the superior Me 109 encountered evenly matched opponents.

For twenty years approximately 33,000 Messerschmitt Me 109s were built in various countries. In February 1937 the first Messerschmitt unit was established in the German *Luftwaffe*, and in November 1965 the last operational unit of the *Ejército del Aire* in Spain was dissolved.

352 air victories were credited to Erich Hartmann, most likely the most successful *Jagdflieger* of all time. Messerschmitt's concept evolved from a light fighter aircraft to a heavily armed and armored utility aircraft. The Czech post-war version C-199 was named Mezec, or mule, due to its horrible flight qualities, and a Canadian pilot who in 1948 took part in the Israeli war of independence in a C-199 thought "they were pieces of shit."

At the beginning of 1939 the Swiss *Flugwaffe* for the first time could, with the Me 109 D, take an all-metal low wing aircraft with retractable landing gear into operation. With the Me 109 E the *Flugwaffe* had a powerful weapon at their command at a critical time. The determination of the pilots to rigorously secure the Swiss airspace in the Spring of 1940 had a large impact on the defense of the entire nation.

The implementation of the Me 109 G led to a short but rather turbulent period with the Messerschmitt aircraft. Although full of technical flaws, the twelve aircraft were the only modern interceptors of the Swiss *Flugwaffe* as of Spring 1944.

Literature on the Me 109 is complicated, and provides little clarity on this interesting aircraft. Over the decades it was often written off, copied, and interpreted. On one hand, this led to portrayals that did not withstand any accurate testing, and on the other hand, much that was worth knowing was never mentioned.

As I began my investigation of the Me 109 in Switzerland at the end of the '80s, an abundance of documents became evident piece-by-piece, that up to now were never analyzed and evaluated. Today, ten years later, the work is here. However, it would be pointless to claim that the history of the Swiss Me 109 would be written 100% accurately. The author gladly accepts sound tips and suggestions.

The publication here does not only wish to describe the operation and technical aspects of this aircraft, but also call attention to the entire problematic nature of the aircraft's equipment and sourcing of material at this time. Thus, the incompetence, bad planning, and internal disputes should not remain concealed. The informed reader will be able to determine the parallels to the present time.

It lies in the interest of the author to recognize our fathers and grandfathers; in a difficult time they succeeded in the single trial by fire of our *Flugwaffe*.

Georg Hoch
Payerne, August 1999
July 2004

The forefather: The Me 109 V-1 Work No. 758.

Chapter 1:

The Development of the Fighter Aircraft in the Thirties

At the beginning of the '30s one toiled with the development of fighter aircraft. The influence of the First World War was still clearly recognized: strutted double-deckers with fixed landing gear, tight propellers, open cockpits, and MG armament were standard. Even the classic two-seater fighter aircraft with a MG shooter was pursued further. In marked contrast to this, the development of the civil aircraft continued: cantilever all-metal aircraft in semimonocoque construction, retractable landing gear, variable pitch propellers, landing flaps and more were introduced.

In the year 1931 the fighter aircraft Boeing P-12E of the U.S. Army flew with a 500 PS engine at approximately 300 km/hr. At the same time, the Lockheed Model 9 "Orion" introduced by Swissair reached 350 km/h with a 550 PS engine, and *nota bene* as a four-seater airliner.

It was not until the middle of the '30s that the Boeing P-26 and Dewoitine D.500 all-metal low wing aircraft were used, where the open cockpit and the fixed landing gear were retained. With their engines of the 600 PS Class they reached approximately 350 km/hr. Similar differences between the fighter aircraft emerged during the development of the bombers. One of the first bombers—which was developed after the most recent realizations of aeronautical engineering, and was set into operation in 1935—the Martin B-10 was approximately just as fast as the fighters at this time. At the parade for 1 May 1935 in Moscow, the experts marveled at a fighter aircraft that was introduced by the troop since the end of 1934: the Polikarpow I-16 Model 1, with retractable landing gear and a closed cockpit. These technical innovations, however, were not absolutely popular among the pilots. The landing gear had to be laboriously operated by hand.

With a closed cockpit it proceeded similarly. The pilots felt constricted and limited in vision. Thus, later versions of the I-16 were again delivered with open cockpits. The development in Japan and Italy proceeded similarly. Aircraft such as the Mitsubishi A5M2, Fiat G.50, and Macchi C.200 had to be produced with open cockpits in the series prototype.

For a long time the armament aspect was conservatively contemplated. Until the middle of the '30s two machine guns from 7.5 mm to 7.9 mm caliber were common arms. There were efforts for aircraft cannons already in the First World War that could fire explosive shells. The strong recoil force of such a weapon did not allow for installation in the wings at that time.

In 1912 the Swiss Franz Schneider, then head design engineer at L.V.G. in Berlin, developed an inline engine with four hanging cylinders and a reduction gear that operated above. This aircraft engine, known as the Daimler E4uF, allowed for the first time the installation of a machine gun that could shoot through the propeller hub. However technical problems, especially the lubrication of the hanging cylinders, prevented a series production of this advanced aircraft engine.

At the end of 1916 the French flying ace Georges Guynemer wished for a weapon with which one could fire off explosive shells. Thereupon the Swiss Marc Birkigt, co-founder and head design engineer at Hispano-Suiza, modified his newly developed V8 engine. The result was the SPAD S.XII with a 220 PS-Hispano-Suiza 8C engine and a semiautomatic 37 mm Puteaux-Aresenal cannon. Although very effective, the weapon was not popular among the pilots; they preferred light and rapid firing automatic weapons. The demand for a cannon could be realized as early as 1933.

As an experiment, Birkigt built his legendary 690 PS-HS 12X engine and a 20mm HS S7 cannon in a Morane-Saulnier MS 227. This then led to the moteur-cannon HS 12Xers or HS 76, which was first serially built in the fighter Dewoitine D.501.

The development of the fighter aircraft of the future was predetermined as of 1935. On one hand, it was the aircraft such as the Hawker Hurricane I and Morane-Saulnier MS 405 that embodied the transition from the biplane of a mixed construction to a monoplane of a semimonocoque construction; on the other hand, advanced aircraft such as the Seversky P-35 or the Heinkel He 112. Two aircraft that subsequently wrote history flew in May 1935 and March 1936, respectively, for the first time: the German Messerschmitt Me 109 and the English Supermarine Spitfire.

11

Chapter 2:
The Aircraft Messerschmitt Me 109

Prehistory

The history of the Me 109 began in 1933 with the development of the Me 108.

Until then, Messerschmitt built using the then common mixed or *Fachwerk* construction that consisted of a supporting steel or aluminum framework, and was furnished with a material or wooden paneling. In France Emile Dewoitine designed aircraft using a light alloy- semimonocoque construction, and in the USA Jack Northrop carried out pioneer work in the field of all-metal aircraft. Although people had experience with light alloy aircraft in Germany since the First World War, they were still predominantly building using a mixed construction.

The Me 108 was a four-seater aircraft capable of aerobatics that was used for training and traveling services. At the conception of the Me 108 Messerschmitt utilized the most modern engineering: cantilever low wing aircraft with retractable landing gear in a monocoque construction. Only the rudder areas were covered with material. Automatic Handley Page fore flaps allowed for good stall flight qualities. Experts believe that an aircraft in this class would hardly be designed differently today.

Mainly Hermann Göring, Erhard Milch, and Ernst Udet characterized politics and the development of the German *Luftwaffe* at this time. Udet championed a fast fighter with good climbing capacity and mobility. In 1934 the bidding for a "*Verfolgungs-Jagdeinsitzer*" (V.J.) went to the companies Arado, Heinkel, Messerschmitt, and Focke-Wulf.

As opposed to the competitors, Messerschmitt had a certain advantage: one could fall back on the experiences with the Me 108. Thus, the new fighter also possessed fundamental characteristics of its predecessor. The designs from Arado and Focke-Wulf were not able to match this. However, it was different with Heinkel.

An "Einflieger" for Messerschmitt.

The brothers Walter and Siegfried Günther could relate to their successful He 70 design. The head of the project, Robert Lusser, would present a provisional mock-up of the future Me 109 already in May. In summer 1934 one could adopt the assembly mock-up of the BMW 116, and at the year's end the definitive design prototype was complete.

At the same time, the construction of the Prototyp V1 began under the leadership of the foreman Asam.

The Me 109 V1/D-IABI, with Messerschmitt foreman Hans-Dietrich Knoetzsch at the controls, started its first flight in Augsburg-Haunstetten on May 28, 1935. The irony of fate was that already the first Me 109 had to be equipped with an English engine. Because the intended Jumo 210—instead of the BMW 116—was not available, a Rolls Royce "Kestrel II" with 695 PS was installed.

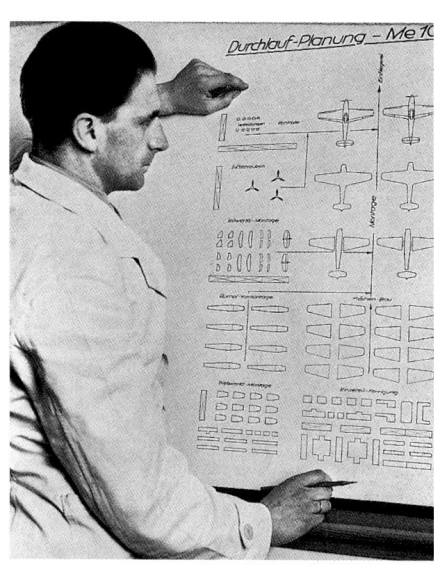

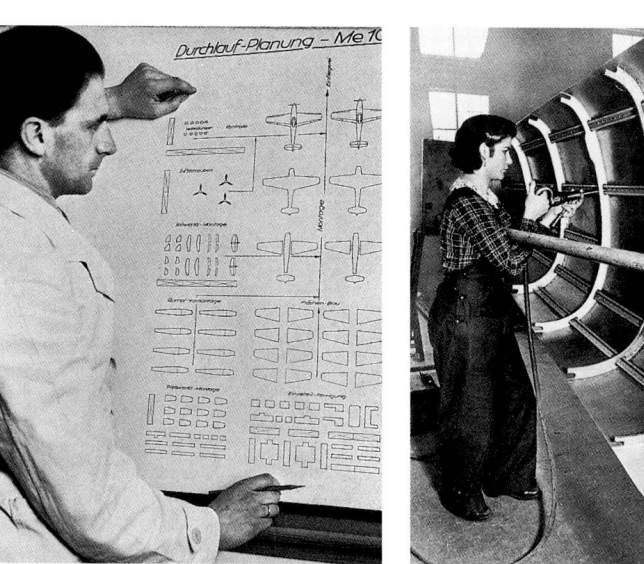

The Messerschmitt Me 109 was already designed for mass production in the planning phase.

Semi-skilled forces could carry out the assembly of the airframe. In the later years of war additional concentration camp prisoners and forced laborers had to perform this work.

The competitive product, the He 112 V1, flew for the first time in September, and was likewise equipped with the Rolls Royce engine.

Heinkel's aircraft was only slightly inferior in performance compared to the Messerschmitt, but was built more robustly, and above all had a wider undercarriage chassis. The narrow undercarriage of the Me 109 was also the weak point of the aircraft. This was already a catastrophe to the factory pilot Knoetzsch on 15 October: at the presentation of the V1 at the Rechlin E-Stelle (testing center) the aircraft crashed during landing, whereby Messerschmitt immediately dismissed Knoetzsch.

The V1 was not available again until February 1936. However, at this period of time the V2 was being tested with the Jumo 210A.

From the beginning, the Messerschmitt design was to be issued for mass production. In order to be able to offset the aircraft's unrigged wings, shock struts were fixed to the fuselage structure, which led to the notoriously small width.

Outwardly, the airframe was designed simply so that unskilled workers could be utilized in production. Even though people leaned moreso towards RLM rather than the He 112, they feared problems in production due to the more costly construction of the aircraft. The Me 109 was not least chosen as the *Standardjäger* for the *Luftwaffe* thanks to Ernst Udet, and the He 112 was released for export. In March and August 1937 the He 112 was also tested in Switzerland.

The Me 109: The Utility Aircraft

The Me 109 proved itself to be very adaptable. The engine power increased from approximately 680 PS to 2000 PS, the weight from 1800 kg to over 3000 kg, and the armament from three MGs of rifle caliber to two 13mm MGs and a 3 cm canon. The aircraft was produced in various countries for more than two decades in more than a dozen series. With a variety of field modification kits "R" and conversion kits "U," the aircraft could adjust to the needs of the troop.

The most important series are mentioned below. From these there were again diverse subvariations and special versions that differed in equipment and in armament. Still, today the experts argue the exact identifier of the single variants. The exact number of designed Me 109s is not known. However, estimations come to approximately 33,000 aircraft.

Aircraft	Engine	PS	MG	Guns	Introduction/Comments
V-1/758	RR Kestrel II	695	-	-	First prototype
V-3/760	Jumo 210C	610	2 MG 17	-	Test Legion Condor
V-13/1050	DB 601 Re III	approx. 1660	-	-	Speed record 1937
B-1/2	Jumo 210D	680	3 MG 17	-	As of February 1937, carburetor engine
C-1	Jumo 210G	700	4 MG 17	-	As of Spring 1938, fuel injection engine
D-1	Jumo 210-D	680	4 MG 17	-	As of Spring 1938, carburetor engine
E-1	DB 601A-1	1050	4 MG 17	-	Winter 1938/39
E-3	DB 601Aa	1100	2 MG 17	2 MG FF	As of Spring 1939
E-4	DB 601Aa	1100	2 MG 17	2 MG FF/M	As of May 1940, Cockpit Armor
E-7/N	DB 601N	1175	2 MG 17	2 MG FF/M	As of August 1940, increased range
T	DB 601N	1175	2 MG 17	2 MG FF/M	For aircraft carrier "Graf Zeppelin"
F-1	DB 601N	1175	2 MG 17	1 MG FF/M	As of October 1940
F-2	DB 601N	1175	2 MG 17	1 MG 151/15	As of January 1941
F-4	DB 601E	1350	2 MG 17	1 MG 151/20	As of June 1941
G-2	DB 605A1	1475	2 MG 17	1 MG 151/20	As of June 1942
G-6	DB 605A1	1475	2 MG 131	1 MG 151/20	As of February 1943
G-6/AS	DB 605AS	1475	2 MG 131	1 MG 151/20	Engine with Höhenlader
G-14	DB 605A1	1475	2 MG 131	1 MG 151/20	As of July 1944
G-14/AS	DB 605AS	1475	2 MG 131	1 MG 151/20	Engine with Höhenlader
G-10	DB 605D	2000	2 MG 131	1 MK 108	As of October 1944
G-10/AS	DB 605AS	1475	2 MG 131	1 MK 108	Engine with Höhenlader
K-4	DB 605D	2000	2 MG 131	1 MK 108	As of October 1944
AVIA C-199	Jumo 211F	1350	-	2 MG 151/20	As of 1947
HA 1109	HS 12Z 17	1800	-	2 HS 404	As of 1951
HA 1112	RR Merlin 500-45	1630	-	2 HS 404/408	As of November 1956

Chapter 3:
Willy Messerschmitt the Person

The statements about Willy Messerschmitt are inconsistent in the literature. Some of his designs, his membership in the NSDAP, as well as the employment of forced laborers in his factories leave some questions open.

His father, Ferdinand Messerschmitt, was born on September 18, 1858, in Bamberg (Bavaria). Although the Messerschmitt family managed a wine business for two generations, he focused on engineering.

From Autumn 1877 until August 1880 he studied at the *Abteilung* III of the Mechanical-Technical school of the Swiss polytechnic college in Zurich.

In January 1883 he married Emma Weil. In the '90s he left Switzerland and his wife in order to live with a girlfriend in Frankfurt am Main. As the second child of this relationship, Wilhelm Emil was born in June 1898. When Willy Messerschmitt grew up Germany was already a powerful nation. The states under Bismarck were united under the Greater German Reich. Names like Daimler, Benz, and Diesel were known throughout the entire world. Graf Zeppelin, with his airship as a concept, was the epitome of aviation. Thus, it was hardly surprising that engineering and less of the wine business appealed to the young Willy Messerschmitt.

In his younger years he came in contact with Friedrich Harth, a design engineer of gliders. Thus, the future path of Messerschmitt was established.

From 1918 until 1923 he studied mechanical engineering at the Technical College in Munich.

His dissertation was the design of the glider S.14. With the S.15 Messerschmitt focused on engine powered flight.

In 1923 in Bamberg he founded his own aircraft factory, the Messerschmitt *Flugzeugbau* GmbH.

He had great success with his five-seater commercial aircraft M 18. In order to secure its production, he merged with the "Bayerische Flugzeugwerke" in Augsburg in 1927.

In 1929 Switzerland bought a BFW M 18c as a surveying aircraft. Because the aircraft had proven itself, in 1935 two further BFW M 18ds were procured.

In 1928 the state of Bavaria withdrew from the BFW. A financial syndicate, together with Willy Messserschmitt, took over the majority of the shares, and the company was privatized. Structural problems with the commercial aircraft BFW M 20 and financial straits practically led to the bankruptcy of the company at the end of the '20s despite full order books. With the takeover of the National Socialists in 1933 production could be completely picked up again. Whether or not an entrepreneur confessed himself to the NS Regime is not a topic of this book.

However, the fact is that Messerschmitt was an active party member: March 1932; "associate" of the SS, Mai 1933; NSDAP member No. 342354, December 1933; Deutsche Arbeiterfront, July 1937; and NS-Bund Deutscher Technik. Further activities followed. Also, the forced employment of convicts and concentration camp prisoners in the Messerschmitt factories must be seen within the spirit of the time.

With the new regime the boom in the armaments industry arrived. Messerschmitt portrayed himself as responsible for an entire series of aircraft that wrote history: the Me 108 travel aircraft; the Jäger Me 109; the Me 209 record aircraft; the Me 163 *Raketenjäger* and the first built in the series; and also the deployed *Jet-Jäger*, the Me 262. The nearly completed prototype P1101 that was captured by U.S. troops in an underground hangar by Oberammergau had a great influence on the development of future *Überschall-Jäger*.

On the other hand, in 1942 Messerschmitt fell out of favor with Milch and Göring due to problems with the Me 210. This resulted

14

in Messerschmitt's deprivation of power within his own affiliated group. Subsequently, Messerschmitt was still responsible for the development and design office.

After the collapse of the Third Reich Messerschmitt came to England with other important German aviation figures. The preparatory work on the subject of transonic flights was of special interest for the Englanders.

In the East and West, German scientists and experts were in demand, no matter if they were accused of any crimes.

In 1948 Willy Messerschmitt was de-Nazified. His activities in the NSDAP were not classified as grave, and also the accusation of being a war criminal was dropped.

Messerschmitt was not even 50 years old, and was full of energy as before. However, his factories were destroyed, and the construction of aircraft was forbidden in Germany. Nevertheless, he established a factory in Augsburg and manufactured prefabricated homes, an undertaking in demand in a destroyed Germany. With his legendary bubble car, in the '50s he scored further successes.

It was merely a question of time when Messerschmitt would design aircraft again. The opportunity arose in 1951 after a visit with General Franco in Spain. Subsequently, the training aircraft HA-100 with a piston engine and the light ground attack and training aircraft HA-200 with two Jet engines emerged.

The project HA-300, a supersonic fighter, began in the summer of 1952. The wind tunnel attempts took place in F+W Emmen. Constant concept changes and the resulting altered deadlines and exceeding costs, as well as internal rivalries prevented the completion of the prototype. The Spaniards attempted to make the best of the financial disaster, and sold the building rights to Egypt.

President Nasser took advantage of this opportunity in order to equip his battered air force with the newest material after the Sinai campaign of 1956. Thus, in 1959 he was able compel Willy Messerschmitt to continue with the project.

The first flight of the prototype HA-300 began on March 7, 1964, in Heluan by Cairo.

With this, Messerschmitt brought upon himself the anger of the Israelis and their affiliated circles. Memories of the Nazi period were evoked once more.

Various problems prolonged the commissioning of the HA-300, and after the "Six-Day War" of 1967 the matter was finally settled. In 1955 the construction ban on aircraft was lifted.

In 1956 Messerschmitt founded the *Flugzeug-Union Süd* with his longtime archrival Ernst Heinkel. In 1969 the later established Messerschmitt-Bölkow GmbH developed into Messerschmitt-Bölkow-Blohm GmbH, or MBB, that in 1989 totaled roughly 40,000 employees. Due to the union of the German airline industry in May 1989, the Giant DASA, Deutsche Aerospace AG, finally emerged with 80,000 employees.

On September 15, 1978, Willy Messerschmitt died at the age of 80 after a difficult operation in a Munich hospital. Leftist groups, however unsuccessful, opposed his burial at the cemetery in Bamberg—his past caught up to him one final time.

Messerschmitt BFW 109, Bf 109, or Me 109?

The question as to how Messerschmitt aircraft are to be correctly named depends on the point in time of the introduction of the particular model.

In 1917, when the "Otto-Werke" in Munich were taken over by the Prussian "Albatros," the company was henceforth called "Bayerische Flugzeugwerke." After the end of the first World War the factory was closed. In 1926, after the takeover of the "*Udet-Flugzeugbau*," the company was newly founded under the name "*Bayerische Flugzeugwerke*," or "BFW." Separate from this, in 1923 Willy Messerschmitt founded the "Messerschmitt *Flugzeugbau* GmbH" in Bamberg. The names of his aircraft began with an "S" or "M." In 1927, with the *Bayerische Flugzeugwerke*, he founded a syndicate. Messerschmitt was to develop future aircraft, and the BFW was to build them. Thus, the aircraft were called "BFW M...."

In 1933 the RLM (*Reichs-Luftfahrts-Ministerium*) distributed new names and numbers for manufacturers and aircraft. From then on Messerschmitt's aircraft were called "Bf." In 1938 the "*Bayerische Flugzeugwerke*" was renamed "Messerschmitt AG," and the identifier was "Me."

Thus, one would have to identify the Messerschmitt 109 before 1938 with "Bf," and afterwards with "Me."

Contemporary documents, such as technical instructions, blue prints, spare parts catalogs, and newspaper articles spoke of "BFW 109," "Bf 109," as well as "Me 109."

The identifiers "BFW" and "Bf" were rarely used in Switzerland. In correspondence and in the troop, the Me 109 D was referred to as "Jumo," and the Me 109 E as "DB," respective of the engine.

More rarely does one find the name "David" and "Emil."

The Me 109 was briefly called "Gustav." In this publication the identifier "Me" will be used.

Chapter 4:
The Me 109 at the International Aviation Meeting in Dübendorf 1937

International Debut of the Me 109

From July 23 to August 1, 1937, the Fourth International Aviation Meeting of Zurich took place in Dübendorf, a mass rally like no one had ever seen in Switzerland. The effort to shape this event to be a great demonstration of aeronautical engineering completely succeeded. Approximately 300 competitors from 13 nations with roughly 160 aircraft attended. 200,000 spectators followed the ten-day event in Dübendorf. Here it is only marginally mentioned that the contemporary press spoke of 10,000 automobiles in the parking lots.

Thereby, the unifying possibilities of neutral Switzerland were once again evident. Already in 1922, at the first International Aviation Meeting of Zurich-Dübendorf, it was possible in Switzerland to bring together former opponents. In 1937, as international tension had taken on threatening measures, one could confront the future opponents once more in a last peaceful competition. Germany

Above: On the evening of July 22, 1937, Udet arrived in Dübendorf with the Me 109 V14/D-ISLU.

Left: In Dübendorf Willy Messerschmitt did not need to fear his competition.

Ernst Udet and his aircraft attracted a great deal of interest. Observe Udet's civilian clothing and white flyer's cap.

The Me 109 B-1, factory number 1062, was flown over to Dübendorf by Dr. Wurster from Freiburg i. Br. on 23 July.

The Me 109 V/9/D-IPLU was used for the international Alps flight in Dreier-Patrouille.

utilized this occasion for an impressive demonstration of technical superiority of its aircraft.

The German team, with approximately 36 aircraft and 90 pilots, attendants, and technical personnel, was absolutely the largest. Familiar names like General Milch, *Generalmajor* Udet, and Hanna Reitsch were also present, as were the design engineers Gerhard Fiseler, Carl Clemens Bücker, Wolf Hirth, Gerd Achgelis, Ernst Heinkel, Claudius Dornier, and, naturally, Willy Messerschmitt. High-tech aircraft like the Me 109, He 112, and Do 17 left one guessing what was to be expected.

Already on June 20, 1937, nine Me 109s participated in the celebratory opening of the Budapest-Budaörs airport in Hungary. Due to strong winds the performance was unable to proceed as planned. Afterwards, the Dübendorf meeting was the first presentation of the Messerschmitt Me 109 at an international mass rally. Officially, the aircraft were announced as "BFW Me 109s with Jumo 210 engines with 640 PS" and "BFW Me 109s with DB 600 engines with 950 PS."

Four or five aircraft were flown over to Dübendorf on the evening of 22 July. One day later Dr. Wurster flew over factory no. 1062 from Freiburg in Breisgau to Dübendorf.

However, several sources speak of a total of six aircraft. For certain, five Me 109s participated in the various competitions.

The Me 109 V7/D-IJHA had a B-0 airframe with a Jumo 210G engine with fuel injection and 730 PS takeoff power, as well as a bigger fuel tank and trunk. The paint used was RLM 63 light gray. The aircraft was predominantly used for tests.

The Me 109 V9/D-IPLU likewise had a B-0 airframe. The aircraft was later equipped with two MG 17s above the engine, and utilized in Rechlin for engine testing.

The Me 109 V13/D-IPKY was a reconstruction from a B airframe with an enhanced fuselage profile to the engine bearer for the Daimler-Benz engine. Officially, the aircraft was announced with a DB 600 engine. Several sources speak of a DB 601. However, verifiable evidence for this is not at the author's disposal.

The D-IPKY was provided in Autumn with an aerodynamically improved engine cowling, canopy, and propeller hub. At this point in time a sophisticated DB 601 Re III with 1660 PS was built in. In this performance the V13, under Dr. Wurster, reached a speed record of 611.05 km/h on November 11, 1937.

Me 109 V7	Factory No. 881 D-IJHA Jumo 210G
Me 109 V9	Factory No. 1056 D-IPLU Jumo 210G
Me 109 V13	Factory No. 1050 D-IPKY DB 600 or 601
Me 109 V14	Factory No. 1029 D-ISLU DB 601
Me 109 B-1	Factory No. 1062 Camouflage Paint Jumo 210

The Me 109 in Dübendorf.

Carl Franke accepts congratulations. Willy Messerschmitt is to the right next to Franke.

The Me 109 V14 was structurally identical to the V13, and had installed a DB 601 against official German instructions. The color of the D-ISLU was kept in a shade of carmine, approximately RAL 3002.

The Factory No. 1062 was a B-1 series aircraft, and had the then common segment camouflage paint RLM 70/71/65. The aircraft was chosen for the weapons installation of three MG 17s, which are seen on the left side of the engine cowling on the exhaust air opening for the engine MG. After the Dübendorf Meeting the Legion Condor, in the Spanish Civil War, deployed the aircraft.

An accurate determination for the second Me 109 B with camouflage remains unclear. Whether the aircraft was utilized in competition or was only exhibited is undocumented. A photograph by the Berlin photographer Inge Stölting at least proves the existence of the aircraft.

No Real Competition for the Messerschmitt 109
The competition regulations in Art. 7 provide the definition of military aircraft. It states "Only such aircraft will be addressed as military aircraft that were verifiably utilized as army property in at least one *Fliegereinheit*, and were flown by military pilots in uniform."

However, this only partially pertained to the Me 109: Germany presented one (or two) Me 109 B with camouflage paint. In fact, the Me 109 B was in action at this point in time with JG 132 and the Legion Condor.

However, the test aircraft were admitted most likely only due to a generous interpretation of the regulations. Some applicants were recognizably angered by it, and forewent participation. Unfortunately the French, who had entered a Morane 405, forewent as well.

While the remaining competitors with series aircraft in a technical class of the early '30s entered, the German pilots with aircraft of the newest generation were ready for the competition. Above all the test aircraft V13 and V14, with the superior Daimler-Benz engines, did not offer the competition any chance. Rather nearly any, because the sophisticated aircraft also had its pitfalls.

The first test was the international speed competition on Sunday, 25 July, in front of over 80,000 spectators.

The race was held on a circuit from the Dübendorf airfield – Wil church tower – Grüningen church tower – Bachtel observation tower – Wangen church tower. The stretch measured 50.5 km, and had to be flown four times, for a total of 202 km. The control station on the Dübendorf airfield had to be flown over from a 30 to 50 m elevation each time.

Only four applicants had entered the competition: a German, two Frenchmen, and an Englishman. The Germans wanted to be certain, and entered a Me 109 V14/D-ISLU equipped with a sophisticated DB 601 motor in the race.

The Me 109 V7/D-IJHA. Notice above the water cooler the additional air intake to the oil cooler. The round wheel shaft and the massive balancing weight for the aileron are noticeable.

18

On the V7/D-IJHA Ing. Carl Franke could opt for the speed competition.

A day later Major Seidemann won the international Alps flight for solo aircraft in the V7/D-IJHA. The aircraft were provided with race numbers for the single competitions.

Both Frenchmen forewent takeoff. The Morane MS 405 that had been entered as a participant was not brought to Dübendorf at all. Officially stated, it was due to technical reasons. Thus, there were two competitors at takeoff: the German Ernst Udet, and the Englander Charles Gardner in a Percival Mew Gull with a 205 PS Gipsy engine. In order to still start the race in an attractive manner, the Englander, in a sporting manner, accepted the late entrance of a second German aircraft with the factory pilot Carl Franke in the Me 109 V7/D-IJHA.

During the first flight around the Bachtel tower Udet could no longer properly handle the throttle valve of the carburetor, and subsequently had to land in Dübendorf with an overheated engine.

Naturally, with his Mew Gull Gardner had no chance against Franke's remaining Me 109, and ended the race in second place.

For the 202 km long stretch Franke needed 29 minutes and 35.2 seconds, which corresponds to an average speed of 409.64 km/h. Gardner received last place with his aircraft; with 34 minutes and 33.8 seconds he was only approximately 60 km/h slower than the Me 109, and this with over 500 PS less engine power.

During the presentation of awards Franke gave up first prize in a comradely manner on behalf of Gardner.

On Monday, 26 July, between 0600 AM and 1000 hours, the participants took off for the international Alps flight of Category A (*Einsitzer*) and B (*Mehrsitzer*). This competition was already held for the fourth time.

Charles Gardner impresses on his 205 PS Mew Gull, but had no chance against Franke's Me 109 V7.

The V14/D-ISLU wore the number 6 for the Alps flight on 26 July. The aircraft was kept in carmine red.
Udet had a preference for red aircraft. As a pilot of the Jasta 4 in the First World War, he flew his Fokker D.VII and SSW D.III predominantly in this color.

19

The stretch Dübendorf – Thun – Bellinzona – Dübendorf encompassed a total of 362 km.

In the *Einsitzer* category seven participants competed: Major Seidemann from Germany on the Me 109 V7 with race number 1, and Ernst Udet on the provisionally repaired V14. Capitaine Robillon from France on a Dewoitine D.510, and with the AVIA B 534 four Czech pilots wanted to try their luck.

The definitive favorite of the race was once again *Generalmajor* Ernst Udet in his red V14/D-ISLU with race number 6. But the flyer hero was again out of luck. Udet started last at 0951 hours in Dübendorf. After the stopover in Thun he had to make an emergency landing around 1100 hours at Uttig-Gut, in Schwäbis by Steffisburg, with an engine malfunction, whereby he caught the overhead contact line of the Burgdorf-Thun train. Udet sustained only minor injuries, but the aircraft suffered a total loss.

For a long time people worried about the Czech Hauptmann Stanislav Engler. He took off with his B 534 around 0800 hours in Thun on the second leg over the Alps to Bellinzona; however, he never arrived. Subsequent to his emergency landing Udet wanted to fly back to Dübendorf with a Me 108. When he heard of the absence of the Czech, he spontaneously searched the side valleys in the Thun region for the missing pilot. Only when low clouds made a continued search impossible the relentless flew back to Dübendorf.

Engler lost his course in the region of Tirano, in Veltlin, from where he reported via telephone around midday.

Before takeoff for the Alps flight a leaky oil tube on Udet's aircraft caused agitation in the German team.
The V14/D-ISLU already had technical problems during the speed competition.

Ernst Udet's emergency landing site on a blossoming field by Steffisburg, Thun.

The V14 suffered a total loss. However, Udet remained uninjured. In many publications the V14 is portrayed in a blue color scheme. The difference in color of the red band of stripes with the swastika in RLM 23 and the carmine red of the aircraft obviously led to misinterpretations.

There was great interest in the Do 17 M V1 in Bellinzona.
This, a combat aircraft certainly not provided with military equipment, was faster than the fighter aircraft of non-German production.

The definitive victor of the competition, with an average speed of 387.4 km/h, was Major Seidemann, followed by three Czechs and the Frenchman. The simultaneous competition of the *Mehrsitzer*, Category B, revealed an interesting comparison. The victor, with an average speed of 375.5 km/h, was General Milch's crew on the Do 17 M V1, which was steered by Major Polte. This was nevertheless approximately 28 km/h faster than the second category *Einsitzer* (Lt. Hlado on a B 534), or around 54 km/hr opposite the *Standardjäger* of the French air force, a Dewoitine D.510.

On the same day in Dübendorf, around 0400 hours the climb-and-dive competition, extremely spectacular for the public, began. This test was being carried out for the first time, and immediately attracted interest.

Ten aircraft from five nations were at the start: a Me 109 and a Hs 123 from Germany, three B 534s from Czechoslovakia, a D.510 from France, two Fairey Fox from Belgium, and two C 35s of the Swiss aviation troop.

Although originally registered, from France the Morane MS 405, and from Germany the Focke Wulf Fw 56 Stösser, with none other than pilot Kurt Tank, were absent. The participants in this competition had to reach an elevation of 3000m above sea level, and then fly over a finishing line 100 to 300 m altitude in the stipu-

Carl Franke's Me 109 V13/D-IPKY was used in the climb-and-dive flight competition.

The Frenchman Dussart had no chance in the climb-and-dive flight competition on a D.510 against Franke's Me 109 V13.
The aircraft belonged to 4ème Escadrille of the GC II/1.

21

The Dreierpatrouille Me 109 V7, Me 109 V9, and Me 109 B in Bellinzona.

lated direction. Most interesting about this test was that not only the engine power for the climb, but also the breaking strength of the aircraft frame for the dive was decisive.

Carle Franke arrived at the competition with the Me 109 V13/D-IPKY. The aircraft was hauled to the start with a tractor. Afterwards, the customary spark plugs were replaced against those with a higher degree of heat and filled with special fuel.

After a steep climb Franke reached a height of 3200 meters only after 1 minute and 45 seconds, and after 2 minutes, 5.7 seconds following a vertical dive with over 600 km/hr he flew over the finish line. The procedure before the start and the climbing capacity indicated a sophisticated engine.

With 2 minutes, 23 seconds the Henschel Hs 123 V5 came in second place. Although the Czechs utilized a special fuel for their B 534 and increased the manifold pressure, they secured third to fifth place.

The Belgian and Swiss pilots did not have a chance on their two-seaters. On his C 35 Major Högger needed 3 minutes, 34.8 seconds. The Frenchman Sgt. Dussart was disqualified due to a low flight over the finish line.

The International Alps flight of Category C, one-seater and *Mehrsitzer* in *Dreierpatrouille*, was held on Thursday, 29 July. This test was also organized for the first time, and brought the participation of six *Patrouillen* and three aircraft from five nations: Me 109 from Germany, AVIA B 534 from Czechoslovakia, Dewoitine D.510 from France, Fairey Fox from Belgium, and two patrols with K+W C 35s from Switzerland.

The Me 109 V7, V9, and B were sent into the competition with race number 4.

This time the German superiority turned out to be somewhat less: with a total time of 58 minutes, 52.7 seconds, the average speed of the Messerschmitts amounted to 374.8, as opposed to Czechoslovakia with 361.2 km/h. The last Swiss patrol on the C 35 required 73 minutes, 39.5 seconds, which corresponded to a speed of 299 km/h.

During the competition of Category A/B on 26 July the Alps had to be crossed in partial clouds, though a cloudless sky dominated. This also explained the minor time difference between the solo and patrol flights.

Next to additional international competitions, air presentations took place daily, during which the Me 109 and He 112 were also flown.

According to rumors, the head of the Italian *Kunstflugstaffel*, Aldo Remondino, with consent from General Milch, could carry out a flight on an Me 109 as the first foreign pilot.

The Swiss Flugwaffe arrived with two patrols to the international Alps flight.
The C 35 did not have a chance; far defeated, they secured last place.

Ernst Udet also came to Bellinzona, however, apart from competition with the Me 108 "Taifun" D-ICNN.

The victor Patrouille of the Alps flight: Oblt F. Schleif, Hptm. W. Restemeier, and Oblt. H. Trautloft.
In the Spanish Civil War Trautloft received the first Me 109 for testing on the Front. In 1945, as Jagdfliegerinspizient Ost, he belonged to "Mutiny of the Jagdflieger," and was removed from his post.

The superiority of the Messerschmitt aircraft was emphasized by the German media, which corresponded to the times then. This had an impact on historians and aircraft enthusiasts even up until today. One may also ask the question as to which aircraft were then compared. Fighter aircraft that were presented in Dübendorf, like the Fiat Cr.32, Hawker Fury, AVIA B 534, or Dewoitine D.27 or D.510 were one to two generations older than the Messerschmitt 109. That a Fairey Fox or C 35, no matter in which competition, never had a chance against the Messerschmitt 109 is by itself understood. Various states forewent participation with the newest flight material possibly due to political reasons.

The INTERAVIA communication No. 456 from August 4, 1937, contained the following written information:

"How much bigger would the prize have to be for organizers and involved parties, if for example during the Alps flight the French Morane 405 or the English Supermarine Spitfire appeared next to the German Bf 109 *Jagdeinsitzer* at the starting line? In the future should not the fear of loss of prestige in the field of aviation be relieved through the Olympic sportsmanship?"

The answer to this was two years in the coming.

The He 112 A-03/D-IZMY was not entered against the Me 109 in the competitions due to inscrutable reasons, which implied that Messerschmitt did not have any real competitors in Dübendorf.

During the emergency landing Udet swept along the overhead contact line of the Burgdorf-Thun train.

The Cause of the Accident of the Me 109 V14/D-ISLU

The abortion of the flight of the Me 109 V14 in Dübendorf, and the emergency landing by Thun provided much to talk about, because Ernst Udet was not just any pilot, but rather the highest decorated flyer (62 air victories) who survived the First World War. As a sport and aerobatics pilot he had international success. In the mid-'20s it was said that he was the first to approach the Allmend by Buochs, the present-day military airfield; however, there is no proof. The spectacular dive flights on his Curtiss Hawk II in Dübendorf from June 1934 also made him popular in Switzerland.

However, his military career in the Third Reich was tragic; as *Generalluftzeugmeister* he was forced to commit suicide in 1941.

The V14 with Factory No. 1029 was not equipped, as was officially reported, with a DB 600, but rather with a DB 601 series No. 161.

Dr. Wurster flew in the aircraft on 18 July in Augsburg. Two further flights are entered in Wurster's flight book for 21 July under military designation. On 22 July, between 0922 and 0931 hours, a flight was mentioned for the first time under the designation D-ISLU. On the same evening Ernst Udet flew the aircraft from Augsburg to Dübendorf.

Udet provided the reason of a sudden increase in oil temperature for the emergency landing by Thun. Traces of oil in the fuselage and within the cockpit allow for the presumption of an oil loss as a cause for the engine malfunction.

Regarding its construction, the V 14 could be seen as a forerunner of the Me 109 E. However, the aircraft did not have the standard DB 601 and no military equipment installed.

Members of the railway had to remove the power cable with alligator shears. The emergency landing took place around 1100 hours. The rail traffic had to be replaced by busses until the evening.

The dismantling work by Thun was visibly undertaken somewhat hurriedly. This led to some problems with the investigation of the cause of the accident.

After the malfunction of the throttle valve lever during the speed competition on 25 July was fixed by members of the Daimler-Benz company, the BFW factory pilot, Dr. Wurster, performed a circa ten minute test flight. According to his information the aircraft, engine, and the instruments display were in order.

An additional inspection of the engine yielded, however, a leaky spot in the oil pressure tube, which was immediately fixed. During the following static test minor leaking appeared again of one drop of oil per two minutes which, however, was considered harmless. Thus, the aircraft could be released for the Alps flight.

At the site of the accident by Thun, which was located approximately 1 km before the starting position, Daimler-Benz employees drained the gas and the damaged oil cooler was dismantled in order to prepare for the transport of the aircraft by truck to Augsburg. A first damage assessment yielded the following:

- The fuselage was broken off approximately 1 meter behind the cockpit.
- The forward section of the fuselage with the landing gear connections was badly deformed, and the fittings partially torn out.
- The right landing gear was broken off.
- The bottom engine cowling with the oil cooler and tubes was torn away to the side.
- Due to the impact, various sections on the engine were deformed or torn out, and the first large end bearing was broken, and had broken through the casing.
- Except for on the ailerons and landing flaps, no greater damages were determined on the wings.

After the arrival at BFW in Augsburg on 28 July the inspection of the engine began immediately. The oil loss was conclusively determined as the cause of the engine malfunction. The oil tank was completely empty, and on the engine there were traces of corrosion and tarnishing.

Because there was no fuse wire found, speculation suggested that the oil supply tube had dissolved. Meanwhile, because during the dismantling work at the scene of the accident this factor was not especially paid attention to, if the tubes were correctly assembled and secure, the cause of the loss of oil cannot be decisively clarified.

Thus, the question remains whether during the repair work on 25 July this aforesaid oil tube was correctly secured.

In some publications guilt for the loss of the D-ISLU is accredited to the pilot Enrst Udet. He is alleged to have been overwhelmed with the "complicated technology of a souped-up high-performance engine." Thus, this may be refuted.

Concerned faces at the site of the accident. By means of the available documents Udet was not blamed for this incident, however.

Chapter 5:
The Procurement of Aircraft in the Thirties

The Long Path to New Aircraft

At the end of the '20s the *Fliegertruppen* found themselves in a desolate condition: a clearly defined assignment of the *Fliegertruppe* still did not exist. Some wanted to see them as an independent military service branch, others as merely support for the field army. The Swiss Parliament was occupied with whether precedence had to be placed on warplanes or reconnaissance aircraft. The technical side of the situation fell beyond the pale, as well. The largest part of the flight material corresponded to the state at the end of the First World War. In 1928 this led to a wide discussion on the condition of the *Fliegertruppe* in the March Session of the National Council. There were over a dozen various types of aircraft in action, of which merely 40% were ready for flight. The internal friction found its peak with the resignation of the *Kommandant* of the *Fliegertruppe*, *Oberstlt.* Müller, who had occupied the office since 1920. On June 1, 1930, his successor became *Oberst* Philipp Bardet, an infantryman.

Since 1925 various foreign aircraft were tested by the KTA and the FPD. Subsequently, the French design engineer Emile Dewoitine and his staff of co-workers were under obligation to Switzerland. The Dewoitine D.27 as a *Jagdeinsitzer*, and the Fokker CV as a utility aircraft was suggested as purchases for the troop. For this purpose, the Swiss councils approved a credit of 20 million francs, spread out over 5 years, for the procurement of aircraft, equipment, and replacement parts. This led to protest assemblies of Socialists and Communists. The request for credit was to subdue the plebiscite. However, on June 4, 1930, the National Council decided in favor of military aviation 117 votes to 47. The Swiss industry could familiarize itself with the newest technology in aircraft construction and secure hundreds of jobs for years.

The situation of the Swiss *Flugwaffe* was influenced by many factors during these years. On the political side, one was under more and more pressure of events since the takeover of the Nazis in Germany in January 1933. On the technical side, it was foreseeable that the D.27 and the CV would no longer suffice in a future conflict in Europe. In a memorandum dated July 11, 1933, the head of the military air service speculates about the future of the *Flugwaffe*. The tasks of the *Flugwaffe* and the style of the aircraft were thus newly defined.

With the Fokker D VII the *Flugwaffe* introduced the first fighter aircraft in the years 1920 to 1929. The last D VIIs were first pulled out of air service in 1938.
In the photograph the No. 621 in Thun, 1922.
In 1921 the aircraft was acquired by the Allied Control Commission, and on November 14, 1922, provided to the troop.

The "military instrument" MA-7 was developed by the K+W Thun in 1924. However, the aircraft did not correspond to the ideas of the troop, and was rejected in 1926.

In the search for a modern fighter aircraft one came in contact with the "Constructions Aéronautiques Dewoitine." A first aircraft, the D. I No. 110, was acquired by Switzerland on August 21, 1925. The F-AHAC was provided with the number 671. In September 1941 the aircraft was written off as outdated and was scrapped in the Lucerne depot.

Together with his design team in 1927 Emile Dewoitine could undertake the KTA at the K+W Thun. During the time the D.27 emerged.
Emile Dewoitine (right) in front of the prototype D.27.01
Thun, May 7, 1928

The D.27 was in service between 1931 and 1940 in the *Jagdflieger* units. Subsequently, they were used in the pilot schools until 1944.
In the photograph the No. 210 in Thun on June 22, 1931

The AC-1 flew for the first time in April 1927. The fighter aircraft developed by private initiative by Alfred Comte stood in competition with the D.27 and did not attract interest at the KTA.
In 1929 the AC-1 was provided with the wings of a D.9 and remained in the inventory of the Fliegertruppe as a single copy until October 1941.

27

The He 112 V5 D-IIZO was tested in Dübendorf from March 13 to April 2, 1937.
The aircraft was later presented at a delegation of the Imperial Japan Marine and sold to Japan in 1938.

One held the opinion that the relationship of two-thirds fighter aircraft to a third utility aircraft was no longer up-to-date, and was intended at a new procurement as half *Einsitzer*, half *Doppelsitzer*. The classic *Jagdeinsitzer*, or so was the opinion, would have lost importance as opposed to the *Mehrsitzer*, because the advantage of speed was only minimal.

This misjudgment was also seen in other countries. Above all, in Great Britain the classic *Jagdzweisitzer* with a tail gunner, such as was the trend in the First World War, was further developed.

In August 1934 it was decided that the number of warplanes would be doubled to 300 in the next years. As a result of diverse studies, three types of aircraft were elected: utility aircraft for reconnaissance and security tasks, *Mehrsitzer* light bombers for ground attack and fighter aircraft for air defense, as well as for ground attack.

Thereby, the aim was to develop the aircraft in Switzerland, or at least to build them in Lizenz. The Swiss proprietary developments, however, were limited to the utility aircraft C35 and C36 through the K+W. The development and procurement of fighter aircraft was obviously classified as less urgent.

On March 1, 1935, Hitler's Greater Germany introduced compulsory military service. Suddenly there stood a powerful *Luftwaffe* with approximately 2000 aircraft that, for years, disguised as airlines, flyer units, or police units, were constructed in secrecy. All across Europe this brought about a massive armament.

During a conference of the head of the military air service on December 18, 1935, the *Fliegeroffiziere* received the assignment to speculate on a future aircraft procurement per 1937.

Indeed, this poll turned out numerically in favor of the *Mehrsitzer*, but the ideas for a new fighter aircraft were clearly defined. Speed 450-500 km/h, time for ascent to 5000 m above sea level 5 minutes, ceiling 10,000 m above sea level, and a radius of action of over 1 ½ hours at full throttle.

For armament two MGs, a motor cannon, and 50 kg bombs as standard were considered. Furthermore, blind flight equipment and radiotelegraphy were proposed. The following models attracted the interest of the *Fliegeroffiziere*:

Morane MS 405
Nieuport Ni 160
Loire 250
Dewoitine D.513
Koolhoven

In the available documents it is not certain which of Koolhoven's aircraft was meant.

The He 112 V7 D-IKIK during a test in Thun from July 31 until August 23, 1937. The aircraft was equipped with a DB 600C with 900 PS and had wings that were enlarged by 20m².
According to some sources the aircraft was later used as a V7/U by Heinkel and v. Braun for test flights with a Walter-built liquid-fueled rocket engine.

The He 112 A-03 D-IZMY was provided with a wooden mock-up in order to determine the visibility conditions during the planned construction of a HS 12Yers. This aircraft was also later utilized for tests with a rocket engine.

The certainly advanced, but somewhat curious FK 55 was possibly conceived on paper at this time. Because promotional writings were found in the archives of the Dübendorf *Fliegermuseum*, it can be assumed that this draft from Holland attracted the interests of the Swiss flyers. At this point in time at Koolhoven one also planned a classic *Jagdzweisitzer*. The FK 52 as a biplane would have meant a step backwards.

It is remarkable that German, Italian, and English aircraft were not discussed. A clear claim was that a 20 mm engine cannon in the future fighter aircraft would have to be installed. However, that meant that there were only few engines to choose from.

At this time the French "moteur-canon" from Hispano-Suiza HS 12Xcrs and the HS 12Ycrs were the ones tested and ready to be installed. The German Jumo 210C with a 20 mm cannon was later tentatively built in the Arado Ar 80 V3.

In mid-October 1936 the *Fliegertruppen* became independent, and the "*Abteilung für Flugwesen und aktiven Luftschutz*" was created. At the same time *Oberst* i Gst Hans Bandi, an artilleryman, was appointed the new head of the *Fliegertruppen*.

Already on 10 November the new service department "*Abteilung für Flugwesen und Fliegerabwehr*" was renamed AFLF for short.

The He 112 with the Hispano Engine

At the end of 1936 the Messerschmitt Me 109 was appointed standard fighter aircraft in the German *Luftwaffe*, and the inferior competitors were released for export.

In February 1937 the *Reichsluftfahrtsministerium* (Ministry of Aviation) of KTA disclosed that one was determined to provide the construction regulations of the Fw 159, Ar 80, and He 112 to Switzerland. The Heinkel He 112, however, was of interest.

From 13 March to 2 April a first test of the aircraft was conducted in Dübendorf. The He 112 V5/D-IIZO was flown by Swiss pilots, and was declared "technically as well as operationally" good.

At the end of July 1937, during the aviation meeting in Dübendorf, the He 112 A-03/D-IZMY was presented. Unfortunately, the aircraft was not placed in any contest, because the He 112 was absolutely comparable to the Me 109 based on performance.

However, the He 112 remained of interest, and from July 31 to August 23, 1937, in Thun, the KTA underwent scrutinizing tests of two aircraft.

The He 112 A-03/D-IZMY with the Factory No. 1957 was already presented at the Dübendorf aviation meeting.

The aircraft was equipped with the Jumo 210Ea Factory No. 40688 with 680 PS. The wings measured 17m².

The D-IZMY was equipped with a Jumo 210Ea. A Hispano-Suiza engine was never installed. K+W Thun, August 1937.

The second aircraft, the He112 V7/D-IKIK, Factory No. 1953, was equipped with a DB 600C with approximately 900 PS. The V7 had the revised fuselage of the B Series with a closed cockpit, and the wings increased to 20m².

An essential point of the testing in Thun was that, with a possible licensed manufacture, the HS 12Ycrs engine, also produced in Switzerland, and a motor cannon were installed. In order to ascertain the visibility conditions for the pilots, a mock-up engine cowling made of wood and materials was added on to the D-IZMY.

The calculated earned value with a Hispano engine was somewhat higher than those flown with the Jumo and DB engine.

However, frequent disturbances and bad weather made a flight test of both planes at the desired multitude impossible. The He 112 proved itself as very failure-prone through diverse material breaches and leaks. The accessibility during the repairs appeared complicated, which resulted in long standing time.

For the possible installation of a Hispano engine, a new construction of the engine mount had to be taken into consideration. The armament variant, with a motor cannon and two wing MGs, could only be accomplished with the A03. Whether or not weapons could be built in the double spar type wings of the V7 had to be first clarified with the Heinkel company.

Furthermore, one had concerns that, due to the weight and the relatively small wheels, the creation of fixed runways was not to be avoided during an all-year operation. In Switzerland one was confronted for the first time with the problems of a modern fighter aircraft. A purchase of four He 112s for further testing was abandoned as a result.

Within the same period of time, in July 1937 a Swiss delegation in France was able to inspect the "meilleur chasseur du monde" termed fighter aircraft Morane-Saulnier MS 405, and the utility aircraft Potez 63. The MS 405 was recommended for further evaluation, while the Potez 63 met with disapproval for the time being.

When, in October 1937 there was still no decision for a new fighter aircraft, the legendary National Council, Gottlieb Duttweiler, began with his activities for the strengthening of the Swiss air defense. The aviation meeting in Dübendorf clearly and distinctly demonstrated the superiority of a modern *Jagdeinsitzer* in comparison with the *Mehrsitzer*. Duttweiler also demanded that a delegation inquire after suitable fighter aircraft in the USA.

In November 1937 Swiss pilots in France could fly the MS 405 and Potez 63. The MS 405 did not comply with all specifications based on performance, but was nevertheless suggested for procurement. The determining factor was that a "moteur-cannon" HS 12Ycrs could be installed.

With the twin-engine Potez 63 one was first in the dark as to its purpose. The aircraft originated from a bidding from October 1934 for a twin-engine "muultiplaces légers de défence" for the French air force. The aircraft was to be built in three versions: as a three-seater *Jagdkommando* aircraft (avion commandement de chasse); as a two-seater *Langstreckenbegleitjäger*; and as a two-seater *Nachtjäger*. Further versions of light bombers and reconnaissance aircraft came later. From similar deliberation the Messerschmitt Me 110 originated in Germany at the same time.

For an explanation of a demand of such utility aircraft one suggested the procurement of two Potez 63.

Suddenly, the procurement of fighter aircraft became a great priority. On December 10, 1937, the Federal Council agreed to the purchase of two MS 405 and two Potez 63. At the same time the EMD was authorized to close a license agreement for the Morane aircraft.

In 1938 the events in Europe precipitated: At the "Führer Conference" in November 1937 one spoke in Germany of the "conquest of new lebensraum through force." The Austria Anschluss and the occupation of Czechoslovakia were the first consequences of this. At the peace conference on September 29, 1938, in Munich, Hitler displayed who would determine European politics in the future.

The condition of the equipment and the readiness of the Swiss military aircraft appeared questionable in mid-1938: 59 obsolete D.27 fighter aircraft, of which 20 were provided with night flying equipment. 16 radio sets were available. 11 aircraft were equipped with a K+W bomb rack that was unable to be utilized due to insufficient reliability. The dangerous spinning qualities of the D.27 led to further limitations in operation. Of the 49 CV utility aircraft, 22 were equipped with radio sets, and likewise had no operational bomb equipment. The same problems existed with the C 35. At this point in time 42 aircraft were delivered. In only 12 aircraft the machine guns satisfactorily functioned, and of the main armament—the 20 mm engine cannon—not a single one was installed. Also, the engines exhibited strong vibrations, and the operation of the rudder was faulty.

On July 13, 1938, this moved the head of aviation, *Div* Bandi, to the following commends at the address of the KTA:

"The war aircraft that our *Flugwaffe* has at their command are partly not nearly operational at full volume. This deficiency of the readiness is ascribed to the delays in the development and delivery of important parts of equipment, such as armament, radio sets, etc., as well as to the execution of modifications that in many cases did not occur with the necessary urgency. The entire issue of material procurement must, in this regard, be declared as absolutely intolerable."

The statements from the head of aviation refer to the aircraft that were partially in action for years.

In the German *Jagdgeschwader*, at this time, Me 109 C-1 equipped with engines with fuel injection were in action in eight *Gruppen*. On 29 July, in England, the first Spitfire I was delivered to the No.19 Squadron. A prelude to future aerial warfare was further shown in the Spanish Civil War by the "Legion Condor" with their Me 109.

The pugnacious National Council, Gottlieb Duttweiler, launched a public initiative in July 1938 for 1,000 aircraft. His wishes did not pull through, but the necessity of modern military flying was no longer put into question. Further parliamentarians created more or less official connections with foreign aircraft manufacturers, which occasionally annoyed the official Bern. Thus, in Spring 1938 an aircraft factory in Switzerland was negotiated with the Glenn-Martin Company in Baltimore, USA, without the knowledge of the Swiss Department of National Economic Affairs and the KTA.

After a spiteful correspondence, the conclusion was reached that the state would not participate in any foreign aircraft factory.

As a most eccentric idea, Zurich engineer F.A. Preiss suggested a "Ramm" aircraft with "an ejectable pilot's seat." His idea—a Blended-Wing-Body aircraft—which was intended as a schooling and public aircraft, was personally addressed to the head EMD, Federal Councilor Minger, on 30 June. Although Preiss affirmed that foreign interest existed, the project was not pursued.

The delivery of the two Morane MS 405s was noticeably delayed. The French aircraft industry was generally considered as badly organized and inefficient. Nevertheless, the Federal Councilor decided to implement the license agreement with France, and have 80 aircraft of the advanced MS 406 built in Switzerland.

In a writing from 3 June, the KTA requested the EMD to determine if 30-40 aircraft, including engines, but without armament, would be available to purchase at short notice. It was also looked into if Swiss military pilots were permitted to fly the considered aircraft in the USA.

For some time the KTA was in contact with various American companies, and also with the Curtiss-Wright Corporation, regarding an offer for the P-36 fighter aircraft. The French government had ordered a first series of 100 of the Export Version with the name Hawk H-75 A-1 on 13 May. A series of 30 aircraft for Switzerland would have been able to follow as of April 1939. On one hand, in the USA one wanted to acquire aircraft commensurate to the Motion Duttweiler, but on the other hand, one had diverse reservations. The requested 20 mm motor cannon was not installed in the P-36. One also reckoned with a delivery delay of several months for the instrumentation and radio sets, and armament had to be installed in Switzerland. The vast distance from the USA and the different metric system could also have led to further problems. The head of aviation, Bandi, also pointed out that an independent domestic aircraft industry was of vital importance.

Several years later it was shown how correct he was when the USA imposed an arms embargo on neutral Sweden, and banned the already ordered aircraft.

In summer 1938 officials of KTA and the *Abteilung Flieger und Flab* began to travel frequently. The political situation in Europe rapidly worsened, and the decision for new fighter aircraft was awaited. On 29 July an exposé with the following demands was given to the EMD: an immediate procurement of 40 aircraft abroad. In order to settle the matter, a delegation must travel to Germany, Italy, England, and the USA. In addition to the testing of aircraft for the immediate acquisition, the possibility of license manufacturing must also be examined.

In the process France was not intended, because its domestic aircraft industry could not cover its own requirements.

Dark Clouds above Europe and Still No New Fighter Aircraft
On September 15, 1938, *Oberstdivisionär* Bandi, *Oberstlt* i Gst Ackermann, and *Hptm* Frey from the AFLF, as well as *Hptm* Rüetschi and *Oblt* Wyss from the KTA, traveled to Augsburg to the Bavarian aircraft factories.

The interest of the Swiss was in the "*Jagdeinsitzer* BFW 109" and the "*Jagdzweisitzer* BFW 110."

In the factory, Prof. Messerschmitt personally received them for a short conference.

The briefing on the Me 109 took place on a Me 108 *Taifun*.

The Swiss pilots could afterwards test the Me 109 Jumo and the Me 109 DB in flight. During the late afternoon the flights had to be discontinued due to bad weather. Thereupon a tour of the factory was organized, and aircraft was exhibited on the firing range. The flights could be continued the next morning. Subsequently, a flight demonstration by a factory pilot on a Me 110 took place. In the afternoon, for the return flight to Dübendorf, the company made two Me 108 *Taifuns* available in a generous manner.

In comparison with the Morane 405, the Me 109 with the Jumo 210 fared worse due to minimal engine power. The performance of the Me 109 with the DB 601 engine, however, was addressed as "paramount," which one could expect at this point in time. Solely the high control pressures, and with it the constraint of mobility were conceived as disadvantageous. Also, in comparison with the He 112, the flight characteristics were classified as significantly better.

The performance of the Me 110 was compared to the Potez 63 ordered from Switzerland, and the Breda 88 from Italy: At this time the Me 110 B-1 was produced in series with the Jumo 210G. The air display must have impressed, because Bandi wrote in his travelogue, "with this Germany has at their command probably the so far best known aircraft of this class." Thereby, the capacity for a "great twin-engine aircraft" was especially mentioned.

However, it was the Swiss of all the *Jagdflieger* who exposed the weaknesses of the Me 110 in Spring 1940. This was again later confirmed during the battle for England from the RAF, and in the air defense of the Reich by the USAAF: the "heavy *Tagjäger*" Me 110s themselves needed fighter protection, and suffered great losses.

On September 20, 1938, the head of the *Fliegertruppen*, *Oberstdivisionär* Bandi, requested the immediate procurement of 10 Me 109 Jumos by the end of 1938, and 30 more Me 109 DBs for the beginning of 1939. If the delivery of the Morane 406s was delayed, a further procurement of Me 109s would be provided. Additionally, "several BFW 108s for training and retraining purposes" were considered. However, the procurement of the Me 108 was made contingent upon the testing underway of Nardi training aircraft. The Me 110 was not mentioned in the procurement proposal.

The Messerschmitt factory in Augsburg left behind a lasting impression on the visitors. The aircraft factory that was constructed according to the most modern standards was especially mentioned. Furthermore, the high manufacturing quality of the aircraft was elevated. It is interesting that the factories in Augsburg primarily served the development and the introduction of the serial production, while the mass production took place in branch factories.

From 20 to 22 September the gentlemen Lang, Ackermann, Rüetschi, and Frey traveled to Italy in order to settle a possible procurement of fighter, utility, and training aircraft. The first destination was Turin, in order to inspect the fighter aircraft Fiat G.50. The flight qualities of the G.50 compared to those of the MS 405, and the Me 109 were classified as significantly worse. On the second day the Swiss pilots were able to fly the G.50 for themselves. The negative impression from the previous day was confirmed: the Fiat showed unsatisfactory qualities, and proved to be unreliable, with "a rather malicious character." Furthermore, the long start and landing distances, as well as the bad visibility conditions in the cockpit, were objectionable. The commission decided to not recommend these aircraft.

During the visit of the Fiat aircraft engine factory, the modern facilities and the cleanliness in the workshops were especially emphasized.

The Swiss had the opportunity to talk with pilots from the Spanish Civil War. It was interesting that the Italians in aircraft development predicted a declining trend: in the future, high speed could be done without in favor of climbing capacity and mobility. Only so-called "*Verfolgungsflugzeuge*" against bombers were seen as an exception.

On 22 September, a visit to Breda in Milan was announced. It pertained to the manufacture of the Ba.65 and Ba.88. The utility aircraft Breda 65 was already tested in November 1937 in Dübendorf, and then seen as a possible successor of the C 35. The twin-engine Breda 88 held the same interest as the Potez 63 and the Me 110.

In 1937, the prototype of the Ba.88 reached several long-distance speed records under head pilot Furio Niclot. The fascist regime in Italy created a propaganda triumph out of it. When the Ba.88, complying with their future task as "Aeroplano di combattimentio," was provided with military equipment, the flight qualities significantly worsened. The Swiss Delegation also had this impression during the air display. Head pilot Niclot needed approximately 400 meters until the aircraft lifted off from the ground, and 500-600 meters more in order to win a couple of meters in altitude. In order to fly a full circle the Breda needed a good 40 seconds, which was approximately double what the Fiat G.50 required. Because there was no aircraft with double controls available, the Swiss could not test the Ba.88 in flight.

The impression after the flight display was negative. One was of the opinion that the Ba.88 could only be put into action for a limited time on the Swiss airfields.

Also, for the assignment as heavy *Jäger,* the Ba.88 was seen as unfit. Even the people of Breda were of the same opinion. How correct the experts were with their assessment of the aircraft was seen in war missions. The Breda 88 was of only limited military worth, and was pulled from the units in September 1940.

The training aircraft Nardi 305 was better evaluated. The aircraft was said to have the flight qualities of a modern fighter aircraft. The retractable landing gear, the variable pitch propeller and the landing flaps were seen as suitable equipment for retraining on the new fighter aircraft.

Future Swiss pilots used the Nardi as a preliminary phase. For the Messerschmitt pilots, the Nardi was intended for the transition to the Me 108 and Me 109.

The proposal of the delegation stated: "that one acquires several such aircraft for the time being on which to train the qualified *Jagdflieger* before these same ones arrive at the fast fighter aircraft." In effect, two Nardi F.N.315s, derivatives of the F.N.305, were procured.

Already on September 27/28, 1938, the same delegation visited England in order to inspect the Hawker Hurricane, Hawker Henley, and the Fairey P.4/34.

The Henley and the P.4/34 were considered possible successors of the C 35. Due to the politically charged situation, however, the Air Ministry denied a tour and demonstration of the aircraft. However, most likely to the surprise of the delegation, the prospect of a test of the Spitfire Mk I was presented.

In London, a nervous atmosphere reigned. In public facilities shelters were excavated, and anti-aircraft guns, searchlights, and sound detecting devices were positioned on bridges and squares.

The next day the delegation traveled to Southampton to the Vickers-Supermarine factories.

After a tour of the factory the Spitfire was presented. On a jacked aircraft one can familiarize oneself with the landing flaps and the retractable landing gear. During the afternoon *Oberstlt* Ackermann and *Hptm* Frey tested a Spitfire Mk I in flight. The minimal area loading, and the associated mobility of the aircraft, was especially emphasized. A rough course for the Merlin II was attributed to the rigid two-blade propeller. There were evidently no existing aircraft with a triple-blade variable pitch propeller. The Spitfire left behind a good overall impression, but in the evaluation it was placed after the Me 109.

During the journey home to Switzerland, the delegation visited the Hispano-Suiza factories by Paris. One wanted to examine the progression of the MS 406, which was primarily contingent on the engine and its aerodynamic line management.

However, attention had to be called to the poor working discipline of the French. The level of efficiency in the French factories was declared a third to that of a German aircraft factory.

From October 22 until November 27, 1938, a delegation of the KTA and the AFLS commenced again on a study trip to the USA. American aircraft factories were visited, and the Swiss *Flugwaffe* tested suitable aircraft. In the meantime, the procurement of Me 109s from Germany was a done deal, and this visit, at least what concerned the fighter aircraft, was only of an informative character. Nevertheless, in the Curtiss factories, the P-36 could be flown by the gentlemen Ackermann, Burkhard, and Rüetschi. Furthermore, the utility aircraft Douglas 8A and Vought V-156, which were viewed as potential successors of the C 35, were of interest.

The American aircraft industry was consistently positively evaluated. The performance of the fighter aircraft in production was compared to the European models, such as the Morane 406, Fiat G.50, or the Me 109 Jumo. Capable aircraft, such as the Spitfire Mk I and Me 109 DB were, however, only in testing as prototypes.

The first Messerschmitt aircraft in Switzerland were the BFW M 18. In 1929 the Swiss Confederation acquired a M 18c, and in 1935 two M 18d as surveying aircraft. The M 18d No. 713 was taken out of service in 1954 due to obsolescence.

Chapter 6:

The Procurement of the Me 109 D/E

Hermann Göring and the *Stag Hunting*
Oberstdivisionär Bandi was the driving force for the procurement of the Messerschmitt Me 109. Also *Oberst* Fierz, head of the KTA, "fundamentally agreed with the acquisition of the BFW aircraft according to the present proposal." However, there was also opposition to the purchase of the aircraft. On September 11, 1938, four days before the Swiss delegation traveled to Augsburg, the first Morane 406H was delivered to Switzerland from France, and the preparations for licensed manufacture began.

The aircraft procurement, and dealing with Nazi Germany was politically controversial. The Ministry of Aviation in Berlin hesitated, as well. Göring obviously did not agree with the sale. After all, the aircraft were a product of the newest technology, and the expansion plans of the Greater German Reich ultimately required the armaments industry.

An EMD writing from September 23, 1938, to the Federal Council states: "*Herr Generalfeldmarschall* is currently stag hunting with General Udet. Udet explained, thereby having the best opportunity to receive the approval of the *Generalfeldmarschall*."

Generalluftzeugmeister Ernst Udet, known as a friend of the Swiss, was an advocate for a delivery to Switzerland from the beginning. Whether the stag hunt, Udet, or both persuaded Göring remains to be seen. At any rate, Switzerland was able to procure the most well-balanced fighter aircraft of this time at short notice and on good terms.

The negotiations with the RLM and the Messerschmitt factories resulted in the following:

"1st stage: delivery of two fighter aircraft of the model with which the German *Jagdfliegerei* is presently equipped, and five per month in November and December. This material should be taken from the series of the German army situated in manufacturing."

The J-382 in the Messerschmitt Regensburg assembly hangar, Spring 1940.

2nd stage: delivery of 30 fighter aircraft, 10 per month in March, April, and May 1939 of the very latest model that is now being manufactured in a large series, and during the next months will successively come to the troop. Also, these aircraft should come from the running series.

All of these aircraft would be delivered entirely equipped, including a radio set but without armament, which would be built by us."

This can be read in Exposé No 22.20.5/50 of the KTA to the EMD on October 11, 1938. However, it is unknown why the radio sets were delivered only in the Me 109 D.

Mass production of Me 109 E in the Regensburg factory, Spring 1940.

The J-355 for the Swiss *Flugwaffe* on the factory airfield in Regensburg, October or November 1939. (See the color photograph on page 196)

A completely equipped aircraft, including armament, cost 280,000 francs. Four forty aircraft, this amounted to 11,200,000 francs.

Added to this amount was still 30 percent for replacement and spare parts, including spare engines.

It is striking that, according to information, there is no difference found in price between the Jumo and the Daimler-Benz aircraft. For financial transactions, an agreement was included regarding the delivery of war material in 1937, which was made between the trade department of the Swiss Federal Department of Economic Affairs, and the German Reich's Department of Trade and Industry. According to the special agreement, 50% of the payment was made through the delivery of iron ore from Gonzen and from the Fricktal.

The proposal on the procurement of two Me 109s and Nardi 305/315s occurred at the same time.

The EMD formulated the proposal to the procurement of aircraft, and presented it to the *Bundesrat*. At the "Meeting of the Swiss *Bundesrat*" on Friday, October 21, 1938, the proposal was sanctioned, and the EMD authorized the contract with the Bavarian aircraft factories. The total cost of 14,600,000 francs was charged to the "Credit for the Strengthening of the National Defense."

The delivery of the aircraft was delayed by approximately two months. The delivery of the Me 109 D occurred thereafter within four weeks. After the delivery of the first Me 109 Es (J-314) on May 7, 1939, the message surprisingly came that the delivery of further aircraft would be delayed. The reason for this was the multiple stresses on the wings. Wing deformities and fractures continued to remain a problem for the Me 109. The deliveries proceeded again as of 20 June, and concluded at the beginning of July.

The Controversial Procurement of the Second Series
Because the political situation in Europe increasingly worsened, and a delay in the licensed manufacture of the Morane aircraft was foreseen, the procurement of further Messerschmitt aircraft was attempted.

On March 21, 1939, the KTA was in touch with the Messerschmitt company, regarding a delivery of ten further Me 108s and thirty Me 109s. For the first time a procurement of forty Me 110s was planned. While there existed no objection from the German side to the delivery of Me 108s and Me 110s, there was a notification concerning the Me 110s, that the aircraft were still unreleased for export, and a possible delivery would depend on the "highest of all positions." The proposal from *Div.* Bandi to the EMD on 27 May stated thirty plus twenty Me 109 Es, because one wanted to procure an additional twenty Me 109 DBs for the planned *Überwachungsgeschwader*.

The *Einfliegerhalle* in Regensburg, Spring 1940. In the foreground is J-388.

On 28 June the KTA surprisingly intervened with a palette of critical questions concerning the Me 109. It demanded an explanation from the EMD of the occurrences during air service, and asked questions pertaining to the maintenance and repairs, as well as the payment of the aircraft.

The KTA was of the opinion that the Me 109 would be only partially satisfactory in operation. Numerous technical dysfunctions and defects in construction provided reasons for criticism.

The selection of material and the performance of the Me 109 D were criticized as "not very careful." In fact, parts on the landing gear had to be replaced because, as was subsequently admitted by Messerschmitt, the construction was inadequately achieved. Later, it was learned from the proper authority that, due to the defects determined in Switzerland, changes were carried out in the running manufacturing of over 100 aircraft at Messerschmitt. The Me 109 E was criticized due to its fixed construction, especially the wings.

Additionally, one wanted to gain experience with the aircraft on various airfields in order to better assess the narrow undercarriage. Because the original demand for an engine cannon with the Messerschmitt aircraft was not fulfilled, the question of armament was again raised.

It was further criticized that German locations were very conservative with the publishing of technical handbooks and repair instructions. The independence for which Switzerland strived from foreign countries was not realized.

There were also scruples with a premature conclusion of a contract, for there was a deal of a new trade agreement with Germany. The existing Clearing Agreement, and the special agreement on payment transactions with reference to war materials ran out at the end of June. The negotiations with the Reich's government were evidently exceedingly difficult. It was feared that, with an aircraft purchase at this time, opposing the negotiating partner, great concessions would have to be made.

The answer to these questions came personally from *Div* Bandi on July 16, 1939. He was of the opinion that the Messerschmitts would have proven themselves in operation, because to date approximately 630 hours on the Jumo, and approximately 50 hours on the DB had been flown—a remarkable number of hours in this short span of time. Since the introduction of the aircraft, the incidents were seen primarily as assembly or pilot mistakes, and could not be attributed to the aircraft. The replacement of parts on the landing gear was considered a single noteworthy incident. In comparison to the models, such as the D.27, CV, and C 35, the Me 109, as far as defects and failures were concerned, was judged quite better.

Thus, attention was especially called to the problems with the HS 77 engine that resulted in the removal and dismantlement of the entire C 35 *Flotte*.

The demand that one had to gain more experience on the Me 109 could not be met due to the political situation in Europe. In response to this, Bandi stated, "Our experience that we have had with all new aircraft and engine models is that when once the aircraft and engine are in a flawless condition, both are technically outdated." A statement that to this day has not lost its significance.

The *Fliegerchef* most likely had the more convincing argument than the head of the KTA, for on Thursday, July 27, 1939, during the meeting of the *Bundesrat* in Bern, the purchase of 50 Messerschmitt Me 109 DBs was decided. Credit was extracted from the funds "Expansion of the National Defense and the Fight against Unemployment," as well as the "new extraordinary military credit." The costs consisted of the following:

- 50 aircraft completely equipped at 300,000 francs each.
- Spare engines and propellers, spare material and tools for 6,000,000 francs.
- Various insurance and acceptance costs to 1,000,000 francs.

The entire deal amounted to the sum of 22,000,000 francs.

Worth mentioning is the requirement that a transaction could only take place if the necessary documents were delivered, as well, in order to conduct repairs on the airframe and the engines in Switzerland.

Also, the aircraft of the second series were, in every sense of the phrase, not completely equipped: above all, the decision to procure the radio equipment from France and install Swiss built weapons had negative consequences on the capabilities of the aircraft in war.

On the economic side, there were evidently certain conceptions and demands on both sides that were not sufficiently discussed: of the 22 million francs, 30 percent were in cash, and 70 percent settled by means of delivery of Swiss export goods. At the time of the conclusion of the agreement the sum was, as a result of a better clearing quotation of 9.2 million Reichsmark, still 16.4 million Swiss francs. Within thirty days after the agreement went into effect the KTA had to afford a payment of 30 percent of the total sum. The condition was that Messerschmitt would deposit the sum in a bank in Switzerland approved by the KTA. This was a type of guarantee if unforeseen events interrupted the delivery of the aircraft. However, Messerschmitt began with the delivery of the aircraft without having received payment, nor any invoice.

From 70 percent of the total sum of circa 11.4 million francs, 7.9 million francs were issued for machine tools and instruments, copper wire rod, bronze wire scraps, décolletage articles made of non-ferrous metal, aluminum, and scrap iron, as well as tires for passenger cars and trucks. The remaining 3.5 million francs were intended for payments to other companies that could deliver important goods.

The Delivery of the Me 109 – The Bone of Contention between Switzerland and the German Reich
At the beginning of November 1939, when war reigned in Europe, the delivery was interrupted for the first time after 15 aircraft. Inquiries revealed that in Friedrichshafen, four aircraft that were ready to be collected were denied access by the German government. Berlin had ordered the block on the grounds that Switzerland did not meet their obligations regarding the compensation transactions. Namely, the Maag company in Zürich did not deliver the ordered machine tools. Because their aircraft were urgently needed due to the state of war in Europe, the matter was handled at the highest level. It turned out that the delay of the deliveries from the Maag company was largely affected by preceding orders from Germany that were only partially paid, and that the responsible offices in Berlin were not in a position to render clear and perfect decisions on the matter of payments. Furthermore, Maag asserted that for the production of the entire delivery program, approximately 200-300 skilled laborers had to be dispensed from military service.

As of mid-December, the deliveries of ten aircraft were taken up again for the time being.

On January 11, 1940, a message came from Berlin via telephone that the delivery of the aircraft was again blockaded. According to the Messerschmitt company, ten aircraft were ready in Friedrichshafen, and eight further aircraft were being transported by rail between Augsburg and Friedrichshafen. Berlin justified the block with the failure to deliver automobile tires.

As a barter transaction, an immediate delivery of passenger car tires, as well as truck tires, was scheduled within three to four months. The deal amounted to one million francs. The agreement was signed on September 27, 1939, but was delayed due to administrative issues. The companies Firestone in Pratteln and Huber in

A bad photograph, but a unique document: the Zeppelin hangar in Friedrichshafen.
No less than 19 Me 109 Es for the Swiss *Flugwaffe* are visible. The photograph originated most likely between January and April 1940, when Berlin suspended the delivery of 25 aircraft for Switzerland.

Pfäffikon refused to act as suppliers opposite Germany for political trade reasons. The trade department asked the KTA to undertake this function.

Due to the state of war, the rubber factories could no longer freely command their stock, because the deliveries from overseas were imposed with a ban on exports.

In Bern, one was of the opinion that Berlin reacted in a petty manner. From Germany there were still 25 aircraft valued at over 8 million francs pending, and the tires, worth merely 1 million francs, were out of proportion. The *Bundesrat* itself was called in to achieve the release of the aircraft in a diplomatic manner. A series of explanations and justifications followed on the level of ministers and directors, whereby one had to admit that Switzerland also had only partially adhered to the original sales terms of the delivery of bronze wire scraps and copper wire rods.

In an abstract, inconsequential matter, this entailed that the aircraft deliveries could not begin again until April 4, 1940, and conclude on 27 April, with the delivery of the last two aircraft.

In this regard, *Bundesrat* Motta from the Swiss Political Department reached his colleges of the EMD with the question of whether or not the procurement of Italian aircraft could be contemplated again.

It can hardly be determined today whether these delays had military explanations. The fact is that the German *Luftwaffe* in the Poland campaign had to accept relatively large losses with over 280 aircraft, 67 of which were Me 109s, and also the continuously lost aircraft against the Allies in the west in the so-called "*Sitzkrieg*" (Phony War). Additionally, a much higher number had to be written off due to accidents and damages without enemy influence.

In all documents, such as travel books or procurement proposals, there were no exact names, such as Me 109 D-1 or Me 109 E-3, listed. Only indications like "Jumo" or "DB" were used.

Also, the frequently quoted export notation Me 109 E-3a is not found again in any Swiss document.

During this time of the evaluation by the Swiss *Flugwaffe* of a new fighter aircraft in Autumn 1938, the Me 109 C-1, with a 700 PS Jumo 210Ga engine with fuel injection, was the highest power "Jumo" variant. The armament consisted of two 7.92 mm MG 17s over the engine and in the wings. The Me 109 D-1 that was later delivered to Switzerland had the same armament, but with the 680 PS Jumo 210Da carburetor engine already used in the Me 109 B series.

During the same time, the Me 109 E-0 pilot production prototype with the DB 601A engine with 1100 PS and an armament of four MG 17s was in final testing.

Those delivered to Switzerland were the Me 109 E-3 prototype, with the possibility of installing a 20 mm gun in the wings.

A Me 109 E of the 8./JG 51 at the beginning of the war.
The Emil was the superior fighter aircraft over the skies of Europe until Summer 1940.

Chapter 7:

The Weapons of the Me 109 D and E

Ambiguous Perceptions of the Armament
At the procurement of the first Me 109 series, one evidently did not have any conclusive ideas of the armament of the aircraft. For both prototypes, either two MGs above the engine and two wing MGs, or a motor cannon and two wing MGs were planned. Because Switzerland possessed a highly developed weapons industry, it was obvious weapons would be installed. Also, the demand for the installation of the motor cannon FM-K 38 also had to be taken into account. One was uncertain of the installation of a motor cannon, if this was even technically feasible. Theoretically, an installation with the Jumo 210, as well as the DB 601, was possible. Also, in Germany one attempted to take advantage of a central weapon. In 1936 a 20 mm MG C/30L (some sources speak of a MG FF) was installed into the Me 109 V3/D-IOQY equipped with a Jumo 210C. However, strong vibrations that could not be corrected prevented a standard installation of this weapon.

In fact, the 7.92 mm MG 17 was merely used as an engine weapon. Thus, the demand for a weapon that used explosive shells was not fulfilled. All further attempts to utilize a 20 mm cannon as an engine weapon displayed the same negative aspects. There is no verifiable evidence if Me 109 Jumo or E, except for experimental purposes, were ever equipped with motor cannons. Not until the Me 109 F-1, as of October 1940, were such weapons deployed.

In Switzerland one studied the installation of a cannon from W+F, Hispano, or Oerlikon, according to the demand for a motor cannon. After January 1939 the first Me 109 D was available, and the installation possibilities directly on the aircraft could be tested. Because at this time still no Me 109 E had been delivered, the installation on this model could be tested only by means of technical documents for the time being. Further explanations were dealt directly in the Regensburg factory.

The conclusion was reached that the space behind the engine in both models did not allow a weapon to be installed, especially for a suitable cannon that was not available.

On 27 March the EMD was informed that neither in the Jumo, nor in the Daimler-Benz aircraft could the requested motor cannon be installed.

In keeping with these circumstances, three armament possibilities were suggested by the KTA:

2 Fl MG 29s synchronized in the fuselage with 960 shot each

2 Fl MG 29s synchronized in the fuselage with 960 shot each and 2 Fl MG 29s in the wings with 480 shot each

2 Fl MG 29s synchronized in the fuselage with 960 shot each and 2 Oerlikon FF-Ks in the wings with 60 shot each

With the DB the third variant was decided, while for the Jumo, due to differences of opinion between the KTA and the troop, the first as well as the second variant were chosen.

The German Me 109s were equipped with two 7.92 mm fuselage MGs from Rheinmetall. For the Swiss Messerschmitt the 7.45 mm Fl MG 29 from the Swiss weapons manufacturer in Bern was chosen.

This weapon, a development by the Swiss Adolf Furrer, was in action for years in many operations, and fired the GP 11 still in use today.

In a writing on December 13, 1938, to the KTA, the Fa. Messerschmitt confirmed that an installation of the MG 29 in the fuselage "is essentially possible."

As an urgent measure, from spare stocks and outdated aircraft, such as the D.19, D.9, and DH 5, forty weapons were removed. The weapons installation in the wings of the Me 109 D were set back for the time being.

The installation of the MG 29 in the Me 109 D and E as an engine-synchronized weapon involved extensive reconstruction work on the aircraft, and also on the weapon itself.

The installation of weapons from domestic production had doubtless logistical advantages as far as weapons and munitions were concerned. On the other hand, the German MG 17, aside from once a greater caliber, was seen as more modern from the side of the control.

As opposed to the MG 17 the MG 29 was shiftily installed, which followed an expensive new construction of the gun carriage, as well as changes on the ammunition box and on the munitions channel. In contrast to the German construction, this resulted in a reduction of the quantity of ammunition by more than half.

The loading, securing, releasing, and shooting were executed through a specially developed electro-pneumatic device in the MG 17. In the MG 29 this occurred manually with Bowden cables and loading handles.

The control of engine-synchronized weapons took place until the 30s, primarily according to the Constantinesco principle. The Romanian Engineer G. Constantinesco and the British Major C.C. Colley had developed in 1916 a hydraulic control that immediately stretched worldwide. The "CC-Gear" proved itself for a long time when aircraft were equipped with a fixed two-blade propeller. Modern fighter aircraft with

The cabin of the Me 109 D
Wild R-VI aiming device with strut. Under the aiming device the two loading handles for the fuselage MG. In between the knob for the air control of both wing MG. On the lower instrument panel left, the operation handle for the unloading of weapons. The upper handle for both fuselage MGs, the lower for both wing MGs.
In the middle the electric shot counter. The left counter was for both wing MGs, the right counter for both fuselage MGs. Later the designations "L" and "R" were replaced by "Fl." and "R."

adjustable three-blade propellers, however, required shorter engine timing. This was achieved through a drive with a rotating Cardan shaft between the motor and the weapon that also still securely functioned at over 2000 U/min.

In Germany this led to the development of a control with shock wires that ran in a cable housing. The release was more precise, because certain inaccuracies existed in the drive with control shafts due to the various gear wheels and joints. The control of the MG 17 occurred over the mentioned shock wires.

In Switzerland, the already established solution by means of a Cardan drive was decided on. For this purpose, specifically a control head had to be designed for the Me 109 D and Me 109 E (See annex no. 5).

The Weapons of the Me 109 D
For a long time the armament of the Me 109 Jumo was uncertain. The KTA saw these as the "transition aircraft" to the Me 109 E, and considered two engine controlled MGs as sufficient. *Divisionär* Bandi, on the other hand, demanded in a writing from July 3, 1939, to regard the Jumo as war aircraft, and also to intend the wings for an installation of the MG 29. For the time being, however, only the fuselage MGs, or as one said in Switzerland, the pilot MGs, were installed.

The space above the Junkers engine allowed for the synchronization drive of the MG 29 on the weapons underside. The barrel bore distance amounted to 240 mm. The weapons were intended for the backside, with mechanical shot counters that could be read off from the cockpit. Thereby, a bulky protrusion emerged on the rear cowling sheet.

Unfortunately, there are no available documents on the exact form of the protrusion and the changes to the cowling sheet. The protrusions also possibly contained channels for the exhaust air of the weapons.

The right side of the cabin.
Next to the seat the box for the signal rockets (with arrows). The uncovered exit opening for the signal rockets is located in front of the closed circuit breathing apparatus. This device was not utilized in Switzerland.

The German version had two munitions boxes one after the other, with 1,000 cartridges each. Because of the munitions discharge channel of the MG 29, the rear box had to be removed. The forward boxes contained two chambers, to each 480 cartridges. However, this involved a bisection of the ammunition quantity.

Shortly after the delivery as a prototype aircraft for the installation of two MG 29s above the engine, the J-308 was delivered to K+W Thun. The Swiss Wild R-V1 replaced the German Revi 3c aiming device that was originally installed.

The first test of the weapons took place at the beginning of March by the KTA. Further shooting tests took place by the DMP from March 22-25 in Payerne. Thereby, in 12 shooting flights circa 4,000 shots were fired on ground and water targets. The success was equal to the CV and D.27 at 70-80 percent, and was declared sufficient. However, various points must be corrected:

- The reflecting gun sight had to be reinforced due to vibrations.
- The path of the trigger lever on the control column was reduced from 30 mm to 20 mm.
- The bulky bulges on the rear outer casing that emerged through the installation of the MG 29, and the shot counter gave rise to criticism. For one thing, they were not compatible with the highly developed aerodynamics of the aircraft. For another, there were problems with reading off the counter due to light reflections. It was decided to replace the mechanical shot counter with an electric one, and to relocate it to the cockpit. A counter group was installed under the oil temperature gauge. The left counter was for the wing MGs, the right for the fuselage MGs. However, it was perceived as disadvantageous that each two MGs were operated only on one counter.
- Finally, it was also the powder evaporations in the cockpit that bothered the pilots.

The KTA, responsible for the adjustment of these faults, requested the J-308 to be flown over to Thun. However this was not complied with, because the DMP required the aircraft for the installation of wing weapons. It also seemed unclear here who was responsible for what.

The wings of the Me 109 Jumo were planned for the installation of the MG 17. Tests with a so-called weapons wing, on which the 20 mm MG FF could also be installed, took place in Germany in 1937 with the V12 prototype. However, this weapon was first installed as standard in the Me 109 E-3.

In Switzerland the installation of the MG 29 was decided, which demanded some adjustments.

The release was controlled remotely by means of an electromagnet, which activated the trigger bar. The loading and unloading was controlled by a pneumatic double aggregate.

The feeding of the munitions took place through a continuous belt that reached over deflection pulleys in a channel from the wing root to the tip.

The munitions of the German MG 17 amounted to 500 cartridges. The belts used in Switzerland allowed a total of only 418 cartridges.

The long ammunition produced high friction forces, which caused problems with the munitions supply. As a solution, a MG 29 with a recoil amplifier was installed. During the test with the weapon No. 563 of the left wing 3,567 cartridges were fired, and with the weapon No. 564 of the right wing 3943 cartridges. Thereby, there were again problems with the munitions supply. The utilized metal belts did not stand the test due to too much elasticity in the long munitions channels. This could be remedied with belts made of thicker sheets.

On 22 June the AFLF informed the KTA of the results, and at the same time suggested to carry out the installation of the wing MGs on the remaining nine Jumo from the DMP in Dübendorf. Thus, it once again displayed that the collaboration between single authorities was not the best. The KTA wanted to retain the aircraft guns for the Me 109 Es that were being delivered and remarked, as was previously mentioned, that it did not regard the Jumo as a war aircraft. In fact, the wing MGs were only sporadically installed as of November 1939 (see appendix no. 1).

Due to the installation of the wing MGs the aircraft weight increased, and with it the area loading by approximately 2.5 percent. This was seen in a worsening of the curving maneuverability.

The superiority of the Jumo without wing MGs, as opposed to those with weapons, prompted the *Kdt* of the *Fl Kp* 15, *Hptm* Lindecker, to compose a report to the attention of the *Rgt Kdt*. Thereby, attention was brought to a premature breaking of the air stream during flight.

The explanations with the *Armeeflugpark* revealed that the advantage of stronger armaments would have to be accepted at the cost of maneuverability, and the sudden breaking of the air stream was a characteristic of the Me 109.

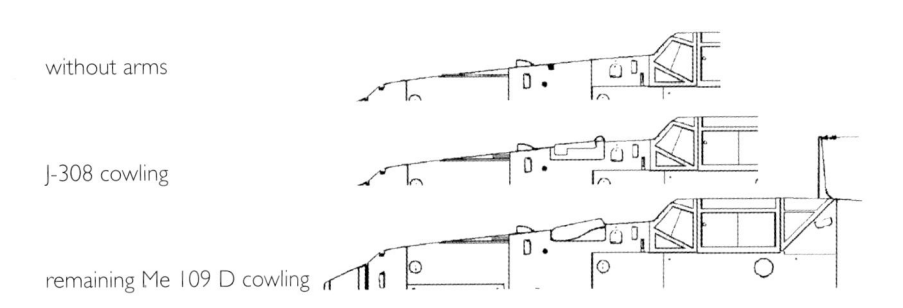

without arms

J-308 cowling

remaining Me 109 D cowling

The Me 109 D J-308 was the prototype aircraft for weapons installation.
Dübendorf, Spring 1939.

The aerodynamically disadvantageous form of the cowling over the MG 29 gave rise to criticism.
Because the weapons of the Jumo were only seldom utilized, a poor firing preparedness was furthermore criticized.

The cowling also possibly enclosed foul-air ducts for the MG 29.
Dübendorf, June 1945.

The original form of the cowling on the J-308 aircraft.
In the rear section it contained a mechanical shot counter. This was later replaced by an electrical counter and relocated to the cabin.
Crash landing in Payerne, September 2, 1947.

The Weapons of the Me 109 E

The installation of the MG 29 as an engine-synchronized weapon had considerable changes as a consequence. The barrel bore distance of both MG 29s was 300 mm, as opposed to 340 mm on the MG 17. This demanded, regarding the design, a sideways shifted drive of the synchronization shaft. The bullet channels had to be correspondingly adjusted.

During the delivery of the aircraft the bullet channels were covered. New cowling sheets replaced the marked protrusions above the controls of the MG 17, and on the rear cowling sheet, a trapezoid-shaped cut with two small protrusions was fit in above the breech casing.

On the eight Me 109s reproduced in Altenrhein, the rear cowling sheet was manufactured of one piece analogous to the German aircraft (J-392–J-399). Likewise, the museum aircraft in Dübendorf (J-355) was subsequently intended for such a cowling.

Just as with the Me 109 D, a single, two-part munitions box with 480 cartridges was used. As opposed to the German original, this involved a reduction from 2000 to 960 shot.

Just after acquisition on May 7, 1939, the J-314 aircraft was delivered to the Dornier factory in Altenrhein as a prototype aircraft for the installation of the MG 29. In contrast with the Me 109 D, the Revi 3c aiming device was reserved.

During firing tests there were defects, especially during the ejection of the shells—a topic that occupied those responsible for a long time.

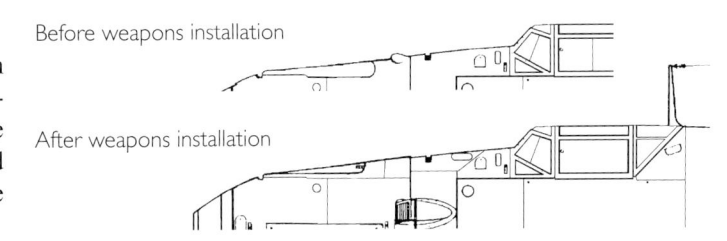

Before weapons installation

After weapons installation

The J-314 crashed a month later into Lake Constance, and was completely destroyed.

As a second prototype aircraft, the J-315 was ready for operation at the end of July for the testing of the weapons. However, during the shooting flights, it was determined that the weapons were incorrectly positioned.

Instead of on the flight axis, they were adjusted on the longitudinal axis of the aircraft. The adjustment on the so-called high-speed flight axis involved further adjustments on the gun carriage of the munitions channels.

On 23 August the J-315 was given over to the *Armeeflugpark* by the KTA for testing within the troop.

During the shooting flights, after several shots the defects already determined on the munitions supply and during the ejection of the shells and belt parts occurred. While firing in a steep flight it resulted in back-ups in the element discharge channel.

The adjustment of these defects lasted until the end of 1939, and limited the readiness for war of the aircraft.

The barrel bore distance was 300 mm, as opposed to 340 mm on the German MG 17.
The MG 29 was installed parallel—the MG 17 offset.

43

At the delivery of the Me 109 E the exit openings for the fuselage MG were covered.
Emergency landing of the J-333 by Zumikon, August 31, 1939.

The cowling sections of the Me 109 E after the weapons installation.
In contrast to the German prototype, the bullet channels were positioned approximately 4 cm closer together. The prominent dents over the control of the MG 17 were removed. Above the breech box of the MG 29 arose two small protrusions.
Emergency landing of the J-374 by Seedorf (FR), September 24, 1945.

Maintenance work on the weapons assembly.
Fl Kp 21, Emmen Summer 1940.

In different countries it was realized that a single armament, exclusively in gun caliber, was no longer sufficient for the demands of a modern aerial warfare.

In Germany, the Me 109 E-1 advanced to the E-3, which was equipped with a 20 mm MG FF in the wings.

Efficient cannons with a high initial velocity, like the HS 404 or the FM-K 38, could not be installed in the wings of a Me 109 due to the large measures. Thus, one had to revert to the more inefficient FF-K (*Flugzeug-Flügel-Kanone*) of the Oerlikon Company.

The effect of the 20 mm munitions was unsatisfactory. The evaluation of the projectile effect on the downed He 111 (see page 97) by Kemleten resulted in an insufficient destructiveness of the explosive shells.

The German *Luftwaffe* also had the same experiences. Not until the introduction of a projectile missile and the MG FF/M could the problem be solved. However, the Swiss *Flugwaffe* did not have this weapon at their disposal.

The Oerlikon FF-K was based on a patent from the German Reinhold Becker from the year 1914. After the Allied victorious powers had forebade every development of aircraft weapons in Germany, Becker was able to connect with the Fa. SEMAG (Seebach Maschinen AG in Zurich-Seebach), and could continue with his work. After the SEMAG had to discontinue production for financial reasons, the assets were taken over by the Oerlikon machine tool factory.

The continued development of this weapon led to a widespread aircraft cannon, and was used in Germany under the name 20 mm MG FF (FF = *Flügel-Fest*).

Weapons up to 20 mm were identified as *Maschinen-Gewehre* in Germany.

The *Bundesrat* decision on October 21, 1938, on the procurement of 30 Me 109 Es was based on the calculation of costs of two MGs and the engine cannon W+F. However, in a duplicate number, the more expensive Oerlikon weapon had to be purchased with the corresponding ammunition. For 60 weapons each with 3600 shot and 10 percent spares a supplementary credit of 2.5 million francs had to be discussed at the meeting of the *Bundesrat* on July 19, 1939.

The aircraft designated for Switzerland corresponded to the E-3 standard, and the wings were already planned for the installation of 20 mm cannons.

The seal of the FF-K was tightened by hand with a lever. The German model had a pneumatic device.

Installation of a munitions barrel. Notice the tight spring sleeve of the FF-K.

The 20 mm FF-K with dismantled cowling sheets. The spring sleeve is slackened. The barrel magazine is not inserted.

The Oerlikon machine tool factory was prepared to transfer the weapons needed on short notice from a running assignment. The original Swiss cannon exhibited several alterations in contrast to the German version. The German weapon was electro-pneumatically controlled. The electrical part served the transmission of *Steuerkommando* and the release. The cocking of the weapon occurred over a compressed air assembly. This permitted the pilot to cock the trigger shortly before the attack, and carry out a reloading movement during misfiring. However, at a drop in pressure or damage to the assembly, the weapon could not be cocked, and consequently fired. In order to avoid this, in Switzerland a device was constructed with which a *Waffenwart* could cock the breech after loading the cannon (see appendix no. 7). During a munitions misfire, however, the weapon could no longer be reloaded in flight. To advance security an additional ignition lock was installed which was to prevent an unwanted trigger of the cannon.

With a switch one could pre-select a single or both weapons (see appendix no. 8).

Thanks to the immediate delivery of the cannons, the installation in the aircraft could begin shortly after their delivery. Initially, there were problems with the shell ejection. Shells flying out hit the rear wall of the discharge box and bounced back into the breech track.

The Oerlikon machine tool factory thereupon modified the shell discharge box, and at the same time communicated with the Messerschmitt company in order to announce these changes in running production, as well as in German aircraft.

The weapons installation in the 50 aircraft of the second series took place as of mid-November 1939.

How arduous the collaboration was between the AFLF and the KTA was shown by means of the already mentioned cannon selector switch. At the Cp av 6 in Avenches, on September 28, 1940, a *Fliegersoldat* released seven shots from the right wing cannon while taxiing with the J-349, although the safety lock was pulled out.

Not until after this incident was it noticed that the lock could have been pulled out at each arbitrary switch setting. The certainty that the assembly was disengaged arose as a result only if the switches were first set on "0," and the key was then withdrawn.

Due to the insecure gun fuse switch, the 20 mm FF-Ks were not loaded as a general rule. During high alert, the munitions barrels were deposited in a tin container under the aircraft.
Fl Abt 3, Avenches, Summer 1940.

The electric spring clips were too weak and had too little spacing. Therefore, an arc could pass over between a spring clip and a live spring, which caused the trigger magnet to respond and led to the release.

The selector switch—a mass product from the automobile industry—was not adequate for the safety requirements of aviation.

Reports, statements, and directives followed. Tests were conducted, and dates were given.

A document was found from the troop liaison officer - KTA dated July 16, 1943, that ends with the following words:

"...the insecure selectors are still installed today after two years and ten months...."

The J-378 on a testing ground in Interlaken. Presumably Winter 1940/1942.
On September 5, 1944, the aircraft was shot down by a P-51 of the USAAF by Neuaffoltern.

The testing of the weapons took place each time after major repairs or revisions. Deployed aircraft had to be tested every three months.

The J-380 on the testing grounds in Payerne, 1945.
The author last used these grounds in the '80s for functional shooting with 30 mm guns ADEN for the Hawker Hunter. The shooting stand is no longer used today for aircraft weapons.

47

The Cockpit of the Me 109 E

Cockpit in the year 1940.
All Swiss Me 109 D and E had a KG 11 control stick at their disposal as standard.
The propeller control lever (under the No. E9) is still located on the middle instrument panel. The gun selector switch (on the lower left instrument panel) is the original version that was susceptible to faults.

The right side of the cockpit.
The single known photograph of a Me 109 E before the installation of the radio equipment.
The map case and the container for the signal cartridge are still installed. The injection pump is located in the original position.

The left side of the cockpit.
The power lever has its original form (without the propeller control switch). At the top right on the chain is the key for the magneto.

48

The cockpit circa 1944.
In place of the propeller control lever are the warning signals for the firearms.
At the bottom left the old gun selector switches, the signal for K-Fuel.

The cockpit circa 1941.
The propeller control switch is located on the power lever. The old position (E9) is covered.
After the installation of the radio, the injection pump was moved analogous to the Me 109 D on the left side.
The signal rocket equipment is not installed.

The Revi 3c aiming device.
In the mid-'30s the so-called reflecting gun sights were introduced, which projected the aiming points on a target. The round gunsight and the ring sight were, however, retained.
The Revi 3c was an early model, and was replaced by the German *Luftwaffe* with the Revi C/12D and 16B. The advanced gyroscopic aiming device EZ42 was used sporadically at the end of the war.
Notice the loading handles of the MG 29, to the top left by the turn and bank indicator, and to the right next to the No. "E9."

49

Tentative installation of a mercury-vapor lamp for the night flight.
Presumably F+W Emmen, Spring 1944.

The Cockpit Me 109 E in its final state. This photograph of the J-355 was taken in the Lucerne Museum of Transport, presumably 1941, where the aircraft was exhibited for years.

On the lower instrument panel of the new gun selector switch, the control unit for the rocket installations and the Morse key.
The control unit for the bomb installations is reconstructed.

The board clock does not correspond to the original model. In the Me 109 E of the Swiss *Flugwaffe* a "Revue Thommen" N-522001 was installed.

Chapter 8:

The Radio Equipment of the Me 109 D and E

Unclear Responsibilities

Broadcasting in the Swiss *Flugwaffe* was under the control of the department for *Genietruppen*. The completely diverse demands of the *Fliegerfunker* were not accommodated. The head of weapons of the *Genietruppen* was of the opinion that all transmissions in the army had to be mandated by a single commando. This not only had a negative influence on the training of the radio operators, but also on the procurement of radio equipment.

Radio technology in the '20s was at the beginning of a rapid development.

Also, in the Swiss *Flugwaffe* tests were performed with radio equipment. These early FT devices (FT = *Funken-Telegraphie*) from Telefunken, AEG, and Marconi were only for A1 operations (soundless telegraphy).

With the introduction of the FG I (FG = *Funk Gerät*) of the Telefunken company around 1930, one could operate for the first time in A2 operation (sounding telegraphy) and A3 operation (telephony). These long wave devices in 300-1300 meter band with a trailing antenna 70 meters long were installed in the DH 5, Potez 25, and CV, and remained in operation until 1938.

In 1935 short wave devices were introduced for the first time with the FG II of the Telefunken company. These devices, which worked in the 50-100 meter band, from then on did no longer require trailing antenna. All CV, and approximately one third of the D.27 were equipped with FG II until circa 1942.

To counter the shortage of radio equipment, in some aircraft the FG II was installed without a transmitter. These receivers, named FG III, allowed the pilots and observers to at least listen to an announcement.

With the introduction of the C 35, in 1937 the Telefunken FG IV was used, with which one could preselect two channels from ground level.

The susceptibility of this radio equipment to disturbances was quite high, which in turn yielded a low operational readiness. Because of the minor number of radio devices, only the *Verbandsführer* often had at his command an aircraft with a complete assembly. In order to optimally adjust the equipment, socalled radio measuring flights took place before the first missions.

In the Swiss *Flugwaffe*, radio equipment was identified as T.S.F. (*Telegraphie Sans Fil* = wireless telegraphy). It was obvious that T.S.F. would be reinterpreted with "*tut selten funktionieren*" (does rarely function).

In 1938 the *Abteilung für Genie* allotted the *Fliegertruppe* the wave range in a 25-50 meter band. The equipment available until then worked, however, in a range of 50 to 100 meters. A reconstruction of the equipment was indeed deliberated, but for the planned procurement of new aircraft new assemblies were considered.

With the procurement of the ten Me 109 Ds, the FG VII from Lorenz used by the German *Luftwaffe* was introduced. These short wave devices, however, worked in a 80-100 meter band, which did not correspond to the directives of the *Abt. für Genie*.

Questionable Equipment Procurement

The search for modern radio equipment for the D 3800 and Me 109 revealed that four models were known or on the market at this time.

Two unspecified devices from Telefunken and Lorenz were available first as a venture, and were therefore not further considered.

In Spring 1939 extensive tests were carried out with the English radio equipment Bell R-10-A. This innovative crystal controlled radio equipment was characterized by simple service and maintenance. The prospect of a delivery was likewise favorable. Because the device could not be used for telegraphy, it was classified as "little appropriate for military purposes," and was disregarded for a procurement.

The device, which caused the crews, radio mechanics, and finally the responsible military officers many worries, was the SIF 450 of the *Société française radioélectrique* (S.F.R.).

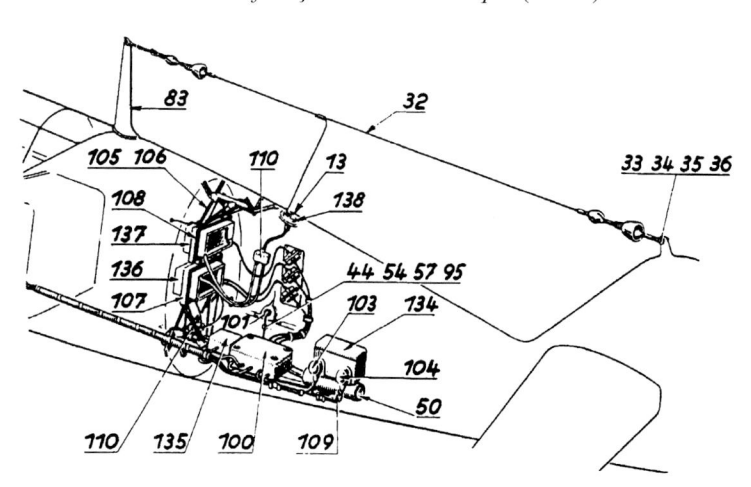

Radio equipment Lorenz FG VII
100 = distribution panel
134 = converter
136 = transmitter
137 = receiver

The SIF 450 was a new design for the French air force, and still not ready for operation.

After inspection, and without having further tested the devices, on July 31, 1939, a procurement of 189 radio sets plus repair material were proposed to the EMD.

For once they all agreed: the *Geniechef*, the *Fliegerchef*, the KTA, and the AFLF.

With this, aircraft radio problems began in Switzerland that persisted until the end of the Second World War.

On August 15, 1939, the EMD authorized the KTA to procure the following material:

1.
– 189 SIF 450 radio sets from the *Société française radio-électrique*:
– 10 devices for Me 109 Jumo (delivered)
– 29 devices for Me 109 DB
1st series (delivered, J-314 in an accident)
– 50 devices for D 3800 (ordered)
– 20 standby equipment

2.
- The necessary spare material for these 189 devices

One device cost 9,300 francs. 92,300 more francs were added to this for the spare material.

The total sum of 1,850,000 was made up with 40 percent compensation assignments for the Swiss industry.

The delivery of the devices, named FG IX, in Switzerland was expected at the end of April 1940. This meant that at the start of war only the ten FG VII radio installations from the Me 109 D were available that, however, did not comply with the wave allocation. Due to the lack of radio equipment, the FG VII continued to be used from time to time.

As a stopgap, the *Armeeflugpark* also changed several ground stations for the installation as aircraft receiver. In order to be able to guide at least single aircraft over the radio, until Summer 1940 36 aircraft were equipped in such a manner. Aircraft that were equipped with radios were often identified with an "R" for radio, or a lightning symbol.

Before the FG IX was available, single aircraft were equipped with a provisional radio set and accordingly marked. Cp av 6, Thun, Spring 1940.

The J-326 was also equipped with a provisional radio set, and marked with an "R" for radio. Most aircraft, however, had no indication of a radio set.
Cp av 6, Thun Spring 1940.

The Problem with the S.F.R. Equipment

In May 1940 the first three devices were delivered from France. The first tests carried out were not satisfactory. The wave meters were so inaccurate that a frequency control between the aircraft and the ground station was exceedingly difficult. In addition, this was the breakdown susceptibility of the device itself.

France's surrender in June 1940 entailed that no further devices could be delivered. The Swiss affiliate of the S.F.R in Bern carried on production. This resulted in a delay in the procurement program. Not until January 1941 could the first thirteen D 3800s be equipped with devices of Swiss production.

In August 1940 it was considered to install the new Telefunken 1005 bF in the Me 109 E. This radio installation, identified as FG X, was planned for the future D 3801 and C 3603. However, the original version was considered the best, and it was definitively decided to equip the Me 109 E with S.F.R.

At the beginning of November the J-377 was provided to the K+W Thun as a prototype aircraft. The installation of the S.F.R. assembly had extensive reconstruction work as a result, which led to delays. The most important changes are listed below.

In the fuselage:
- The mounting of the FG VII radio equipment was planned ex factory on frame 4. The FG IX was installed behind the access panel between frame 5 and 6.
- To enable a vertical bushing of the aerial line to the transmitter, the existing antenna lead in was shifted circa 18 cm forward – The onboard battery was moved into the luggage compartment behind the cockpit. A ventilation opening had to be placed on the deck of the fuselage.

In the cockpit:
- The FG IX control unit was installed on the right side of the cockpit.
- The injection pump, together with the case, had to be shifted to the left side.
- The frequency switch together with the cable pulleys were placed on the right side of the cockpit.
- The signal rocket box and the map holder had to be removed due to lack of space.

Instead of the transmitter, receiver, and both converters, up to four weight plates could be installed.

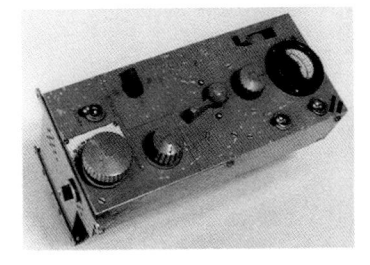

FG IX Operating Device

Radio equipment with problems, the S.F.R. SIF 450 (FG IX)

FG IX Sender

FG IX Receiver

53

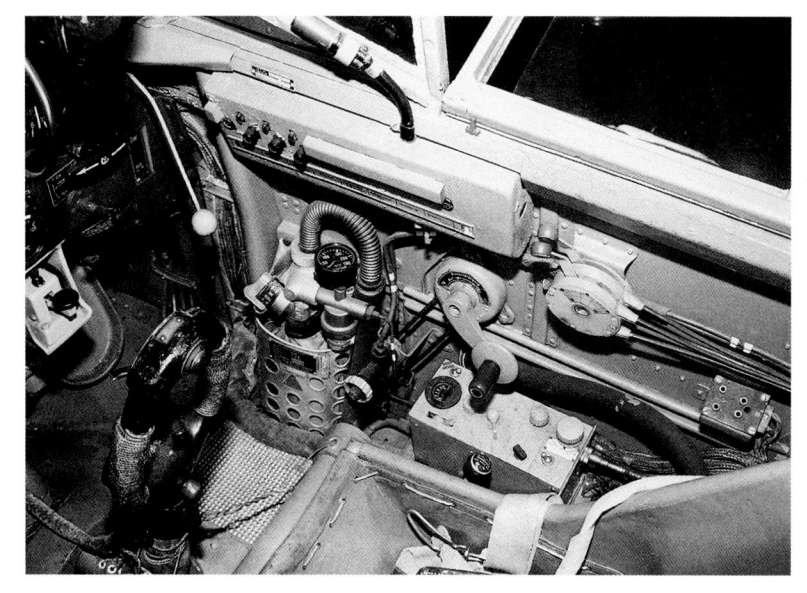

The control unit of the FG IX is located on the right side of the cabin. Both red lines on the upper right corner indicate a normalized device.
Above the device is the wave commutator with two handles. To the far left the Morse key device for the A2 operating mode.
Museum aircraft, Dübendorf.

On January 23, 1941, the model construction could be inspected. Although *Funker-Chef Oberst* Wuhrmann and his colleagues had to object to several points, the installation was accepted as a whole, and the aircraft was flown over to Buochs for the wiring of the assembly. The complicated reconstruction work and adjustment work lasted approximately ten weeks.

On April 4, 1941, the J-377 was the first aircraft equipped with the FG IX flown over to Thun for further testing.

Divisionär Bandi, concerned that the Messerschmitt-*Abfangjäger* still did not have any radio equipment 1 ½ years after the start of the war, demanded from the KTA on February 10, 1941, that, until the end of April, at least 18 aircraft should be provided with radio equipment. The KTA asserted that, due to the extensive reconstruction work, a delivery before July was not possible. Bandi in turn insisted that until 25 April, invasion day of the *Fl Abt* 31 (*Fl Kp* 7, 8, and 9), at least six Me 109 with radio equipment be available. In mid-September 1941 the troop had a total of 22 aircraft with radio equipment at their disposal.

Meanwhile, the first experiences were made with the S.F.R. devices in the D 3800 aircraft.

The technical report from April 22, 1941, of the *Fl Abt* 2, comprised of the *Fl Kp* 3, 4, and 5 with D 3800 (FG IX), as well as the Cp av 6 with the Me 109 E (partly FG VII) turned out surprisingly.

While the transmitting and receiving power of the FG IX was satisfactory, various technical shortcomings to defects in construction were opposed.

On the other hand, the FG VIIs were classified as safe and reliable; only the transmitting power was assessed as somewhat weak.

The installation of the devices took place in F+W Emmen and K+W Thun.

In June the KTA requested that at least three aircraft were to be equipped per week.

Furthermore, it was shown that during the reconstruction in F+W Emmen the frame for the onboard briefcase in the area of effect were riveted together with cover strip of the droppable rear cabin section. This entailed that all aircraft had to be controlled and adjusted if needed.

The technical unreliability of the S.F.R. assemblies generated the saying "*So Funken Russen*." The reports from the retraining courses from June 22-23, 1942, clarify:

Fl Kp 7 (4 Flz)
J-338: fuse burned through, humidity in transmitter converter
J-354: at 7,000 altitude, antenna deflection at 0
J-344: no defect
J-382: no defect

Fl Kp 8 (4 Flz)
1 Flz short-circuit in transmitter converter
1 Flz receiver converter, solder connection broken
1 Flz: short-circuit in the anode circuit of the transmitter
1 Flz: no defect

Fl Kp 9 (4 Flz)
J-316: no heating voltage on the receiver
J-350: connection to the choke broken
J-346: no heating voltage on the receiver
J-378: no anode voltage, connection broken

The manufacturing company could not realize a socalled normalization program conceived by the KTA and *Armeeflugpark* for the improvement of the devices. In Spring 1943 the S.F.R. company was deprived of the assignment, and the Autophon company in Solothurn was assigned the fabrication of the normalized FG IX.

Not until the end of 1943 were all of the radio installations originally ordered from France in 1939 delivered. From the 189 procured devices merely 56 were in air service: 28 in Me 109 and 28 and D 3800 aircraft. The remaining devices were not ready for operation due to repair work or maintenance work.

In order to increase the performance of the FG IX devices, in Spring 1942 a new antenna system was tentatively installed on several aircraft.

On the occasion of a factory flight in Buochs on 16 March, during which the vibrations and stability of the antenna system were tested, the J-340 crashed during landing, whereby the left shock strut was bent.

Another aircraft (J-378) was put into operation on 23 June by the *Fl Kp* 9 for radio measuring flights. The aircraft, labeled as "U-Boot," or "Super S.F.R.," brought good results.

Normally the connection from aircraft to aircraft was poor at short distances, and could completely break off at a distance of several meters. With the J-378, which was located over Valais, could a connection be maintained with a *Patrouille* in the Aarau-Solothurn area without any problems.

However, the raised antenna mast was subject to vibrations and limited the flight qualities. In December test flights were carried out for the purpose of vibration measurements on the antenna mast with the J-321 from K+W Thun. There is no information available on the further utilization of the "Super S.F.R." The aircraft were later again normalized.

The enhanced antenna system on J-340, Buochs, March 1942.

J-340 on March 16, 1942, in Buochs. Notice only the antennas are fixed to the outer wing assembly.

"U-Boot" J-378 of Fl Kp 9.
The measuring flights from June 1942 yielded good results.
However, the high antenna mast was subject to vibrations.

Chapter 9:

Introduction and Operation of the Me 109 D and E

The New Aircraft Me 109 D – New Problems

The Me 109 D-1s for Switzerland were built at Arado Warnemünde, in Lizenz.

Warnemünde—north of Rostock, on the Baltic Sea—since 1913 was the location of naval aviation. In 1917 a branch factory of the Friedrichshafen GmbH aircraft factory emerged. From this factory, in 1925 the Arado trading company arose.

After the Nazis took power, the German aircraft manufacturers were forced to work by order of the Reich's government. Ernst Heinkel, Kurt Tank, or Willy Messerschmitt, all active members, were prepared for unrestricted collaboration with the RLM.

Hugo Junkers or Heinrich Lübbe (principal owner of Arado), however, refused to collaborate with the Nazis. In 1935 this led to the compulsory expropriation and nationalization of the Arado factories. As a "National Socialist model plant," from then on Arado also had to build aircraft and assembly groups for Heinkel, Focke-Wulf, and Messerschmitt.

From the total 647 constructed Me 109 C/Ds, approximately 145 Me 109 D-1s were built at Arado in Warnemünde.

The delivery of the ten aircraft for Switzerland took place from the Messerschmitt factory in Augsburg.

A Jumo of Fl Kp 15.

The camouflage paint RLM 70/65, Swiss national emblem, and the matriculation were already applied in Germany. The aircraft were taken over by pilots of the KTA in Augsburg, and flown over to Dübendorf with a stopover in Friedrichshafen.

The first overflight of a Me 109 from Augsburg to Dübendorf ended with a crash.
The narrow landing gear proved to be a weak point of these aircraft from the beginning.
The J-303 had to go back to the factory, and was not again at the disposal of the Fliegertruppe until the end of May 1939.
Dübendorf, December 17, 1938

The second overflight of a Me 109 D into Switzerland ended in the proximity of Frauenfeld. WIlikon, January 5, 1939

On December 17, 1938, at approximately 1215 hours *Oblt* Mooser took over the J-303 in Augsburg for the overflight to Switzerland. The landing in Dübendorf took place considerably quickly at circa 150 km/h, and after an approximate 110 meter rolling distance the aircraft swerved to the left. After another 15 to 20 meters the right landing gear buckled, and the J-303 crashed. A sudden crosswind was stated as the cause. On January 4, 1939, the aircraft was brought back to Augsburg for repairs, and was not operational again until 30 May. The damage was estimated at approximately 10,000 francs.

With this, the narrow landing gear presented a weak spot of the aircraft from the beginning.

On January 5, 1939, the J-305 was flown over. Shortly after Frauenfeld the water temperature rose to over 120 degrees, and in the cockpit an odor became noticeable. *Oblt* Wyss had to give up the attempt to land in Frauenfeld, and at 1513 hours carried out a belly landing in the open by Ellikon/ZH. The cause was a loose coupling on the water conduit. On 10 January the aircraft was given back to the Messerschmitt factories in Augsburg. The repairs, which lasted until 18 September, resulted in a cost of approximately 50,000 francs.

On 17 January *Hptm* Frey had to make an emergency landing with the J-309 in Dübendorf due to an oil leak. A foreign substance in the oil pump led to engine malfunction. Whether this incident occurred during the ferry flight from Augsburg to Dübendorf cannot be explained free of doubt, due to the contradictory statements in the documents.

There is also a document on the J-310 that leaves questions open: On 5 January the aircraft was taken over. In the flight book of the head of the *Serieneinfliegerei* with Messerschmitt Augsburg, on 17 February there is an entry for a six-minute factory flight on the J-310. It cannot be determined if the aircraft was in Augsburg for repairs.

Further overflights proceeded without incident. However, during training operations the difficulties of a modern aircraft became noticeable.

During a landing on 1 February in Dübendorf, *Oblt* Läderach could not secure the left landing gear on the J-307, despite activated emergency actuation. The consequential belly landing amounted to 10,000 francs in damage. A break of the latch between the working cylinder and the landing gear was the reason for this incident.

After the control flight on 20 April, the J-310 swerved to the right at landing. Thereby, the left wing was so damaged that it was given to repair at the manufacturer's works.

On 31 May Lt Thurnherr had to make an emergency landing in Biel-Bözingen in the J-301 with an overheated engine, because a cylinder stud bolt was broken. The engine (series No. 41 562) was handed over to the Junkers company for warranty repair.

The J-310, shortly after the delivery in January 1939. The aircraft was shot down by Me 110 of the II./ZG 1 on June 4, 1940, by Boécourt.

On February 1, 1939, the J-307 crashed in Dübendorf. Hptm Läderach could not bolt the left landing gear, despite an active emergency release.

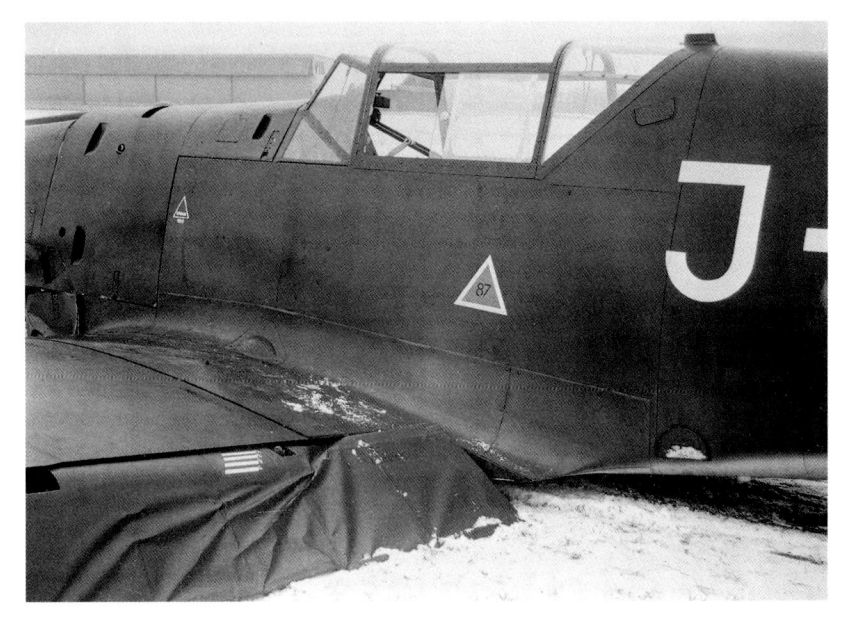

The Me 109 D was operated with gasoline OZ 87 at this time. Notice the jaw teeth clutch for the crank handle on the left side. Also, compare the Carman plates with photographs of the Me 109 E.

The J-307 still had no weapons installed.
Notice the light green border of the filling hole for the coolant, and the brown triangle reference for the lubricant.

The belly landing of J-307 was a spectacular performance. As veterans remember, in the process the aircraft stirred up a giant snow cloud. However, the damage amounted to 10,000 francs.

The retraining on the Me 109 D put demands on the ground crew, as well. If it was not technical problems surrounding the engines, it was the poor starting and landing characteristics, which caused much damage. In the first weeks of retraining alone, five tips had to be replaced or repaired due to contact with the ground. Also, the German *Luftwaffe* complained about the same poor qualities of the Me 109. In the first two months of war, twice as many aircraft were damaged during takeoff and landing accidents than from action with the enemy.

The operational readiness of the Me 109 Jumo was approximately as follows in the first three months:

J-301, J-302, J-306, J-309, J-310: air service with location Dübendorf

J-303: Repairs at Messerschmitt in Augsburg until May 30, 1939
J-305: Repairs at Messerschmitt in Augsburg until September 18, 1939
J-307: Repairs in Dübendorf
J-308: Model aircraft for weapons testing
J-304: No information

With the Me 109 D, the *Flugwaffe* had a modern fighter aircraft at their disposal.

Landing flaps, variable pitch propellers, closed cockpits, and retractable landing gear was a novelty at this time that one had to become accustomed to.

The *Flieger Kp* 15, under *Hptm* Lindecker—equipped with the D.27 up to now—retrained in Spring 1939 on the new aircraft.

For the retraining on the Me 109, the Me 108 was also utilized. Four Taifun were temporarily held by civil enrollment. In the photograph are HB-HEK, HB-HEB, A-205, and A-203.

J-306 in its final descent in Dübendorf.
After an accident in 1942 the fuselage was dismantled, and was later used on Me 109 E/J-392.

Scholastic air service on the Dübendorf airfield. The aircraft are still not equipped with weapons.

The Me 108 and Me 109 D after a cross country flight on the Birsfelden airfield in Spring 1939.

J-314 was delivered on May 7, 1940, as the first "DB" into Switzerland, and was utilized as the model aircraft for weapons installation. Four weeks later it crashed and disappeared into Lake Constance.

Me 109 E: A New Challenge

From six manufacturers, approximately 4,000 Me 109 Es of all versions were built.

The Me 109 E-3 for the Swiss *Flugwaffe* came from ongoing production from the Regensburg factory (Mtt. R). With the aircraft, an export version was included that differed in the equipment from those of the German *Luftwaffe*.

Numerous countries from the Balkans and Eastern Europe, as well as Spain procured the Me 109 E. Japan received approximately ten Me 109 E-3s, and in 1941 Russia received three aircraft for testing. With eighty aircraft, Switzerland was the largest buyer of Me 109 E.

The export label E-3a (a = Ausland (*foreign countries*)) was never mentioned in connection with Switzerland—neither in German, nor in Swiss documents is this identifier found.

The aircraft for Switzerland were not equipped with radio equipment or weapons. Another difference from the German performance was the Revi 3c aiming device and the KG 11 control stick. The camouflage paint, the national emblem, and matriculation were applied in the Regensburg factory.

The delivery of the Me 109 E took place from Friedrichshafen-Löwenthal. The aircraft were partly flown over Regensburg, partly transported via train to Friedrichshafen with dismantled wings. In the old Zeppelin hangar one prepared it for the overflight to Switzerland.

As before with the Me 109 D, pilots of the KTA also flew over the Me 109 Es to Switzerland. However, there are no indications that overflights from German pilots took place directly from Augsburg to Switzerland.

On May 7, 1939, the J-314 was flown over to Altenrhein as the first aircraft. The aircraft was utilized at Dornier as a prototype aircraft for the construction of the MG 29.

Four weeks later on 7 June, the J-314 was flown over to Dübendorf by a KTA pilot for further testing. *Oblt* Suter, who for the first time flew on a "DB," wanted to practice an additional landing approach in Altenrhein. At touch-and-go the right landing gear again failed, and brought the aircraft into a bank. The pilot was possibly blinded by the solar radiation on Lake Constance, as well. The J-314 crashed approximately 100 meters from the shore into the lake, whereby *Oblt* Suter was killed.

Due to the already mentioned shortcomings of the wings, the delivery of further aircraft to Switzerland was delayed until 20 June.

When the delivery of the second series began on October 8, 1939, there was war in Europe. Poland had surrendered at the end of September, and in the West the socalled "Phony War," or "drôle de guerre," reigned.

As is obvious from the dates of acquisition, the aircraft were only partially delivered in sequence of the factory numbers or matriculations.

The experience with the new aircraft indicated that one probably had efficient aircraft, but still had to battle with the same problems:

- The narrow undercarriage led to numerous takeoff and landing accidents.
- Numerous emergency landings as a result of engine malfunctions.
- Material breeches and structural deformations.

The delivery of the Me 109 E:

1st Series	1 Flz	5/7/1939
	29 Flz	6/20 - 7/5/1939
2nd Series	15 Flz	10/8 - 11/7/1939
	10 Flz	12/15 - 12/20/1939
	25 Flz	4/5 - 4/27/1940
Reproduction Doflug	1 Flz	4/28/1944
	8 Flz	7/18/1945 - 3/19/1946

61

The second total loss of a Me 109 occurred on July 14, 1939. Due to an engine malfunction Lt Erwin Wannenmacher from Fl Kp 21 crashed into the grounds of the Netstal Paper Company.

The flight of Lt Wannenmacher of *Fl Kp* 21 proceeded tragically. On July 14, 1939, at 0917 hours, he took off on the J-322 in Dübendorf. Above the Walensee an engine malfunction occurred. Lt Wannenmacher decided to make an emergency landing on the nearby Mollis base. Due to the strong development of smoke, he stopped the engine at approximately 1,000 m/G. The approach of Mollis, however, occurred too high, and forced a turning loop. Thereby, the aircraft lost altitude, and touched the roof of an adjacent building of the Nestal paper factory. The aircraft crashed into the wall at the foot of the main building, whereby four workers were slightly injured.

Lt Wannenmacher was rescued from the wreckage of his completely destroyed Me 109, but was severely injured. The cause for the engine failure was due to a breech in the valve plate on the No. 1 cylinder.

Oblt Streiff, from *Fl Kp* 21, as well, had better luck. Together with *Oblt* Meyner he took off on March 3, 1940, in Dübendorf for a radio communication test flight. Above the Kemptthal, at approximately 100 m/G an abnormal increase of the control pressure was determined. Also, the trimming remained ineffective.

The aircraft abruptly climbed into the clouds, and *Oblt* Streiff had to jump with a parachute. The J-312 crashed by Luckhausen-Ottikon into an orchard, and Streiff landed uninjured in a fir glade.

The examination revealed that the chain break of the stabilo adjustment had been released, and the horizontal stabilizer abruptly shifted to -8 degrees.

The responsible examining magistrate questioned all Messerschmitt units on their experience with the stabilo adjustment. Thereby, it became apparent that as a result of the reduction of the spring tension and material abrasion of the chain assembly, the horizontal stabilizer could independently shift. In a TM (*Technische Mitteilung* = technical communication), the control of the stabilo adjustment was thereupon regulated.

The completely destroyed flight deck of J-322. Lt Wannenmacher survived the emergency landing, badly injured.

The torn Carman plate on J-302. Payerne, March 15, 1940.

Messerschmitt - Lightweight Construction with Defects

The stability of the construction of the Me 109, especially of the wings, gave rise to criticism and speculation.

The English journal "The Aeroplane," wrote in an issue from 1940 that the Me 109 easily "closes it wings" in a nosedive, and generally is not considered one of the best regarding the construction stability. This was simply not accepted in Germany, and the magazine "*Flugsport*" immediately had a counter statement.

The fact is, in 1939 the delivery of the Me 109 E for the Swiss *Flugwaffe* ceased because the wings had to be reinforced. There are no details known to the author what was changed. It is simply remarkable that in contrast to the Me 109 D, the Carman plates on the fuselage side were reinforced with a false edge.

Oblt Rufer, Fl Kp 15, initiated an approach on the Forel shooting range on March 15, 1940. At the start, the wings of the J-302 were highly stressed. An investigation in Payerne revealed that this overloading led to a permanent deformation. At this point the aircraft had 139 operating hours.

Also, after merely 38 operating hours on the J-338, a deformation on the wings' covering was found. One immediately got in touch with the Messerschmitt company, and delivered corresponding photographs and sketches. Messerschmitt in Augsburg was of the opinion that such a deformation could only arise from an excess of multiple weights, as they can appear during interception with great speed. It was calculated at a weight of 85 percent of the collapse load, which corresponded to multiple weights of 9 g.

On 22 and 23 July 1940 at Doflug, in Altenrhein, an expert's report of both wings of the J-302 took place. None other than Professor Amstutz from the *Institut für Flugzeugstatik der ETH* made evaluations. Thereby, one came to the conclusion that the statements from Messerschmitt on the multiple collapse load corresponded with the German building regulations for aircraft of the group 5 (the highest load), but this limit value was set relatively low in Germany. It was held firmly that, for example, in France, for an aircraft like the Me 109, a 23 percent higher value was applied.

The meager stability values were preferred in support of better flight performance in Germany.

Prof. Amstutz further noted: "From the relatively meager stability other, still to this day unrevealed shortcomings could result in time."

The third Me 109 E that went missing was J-312. As a result of the failure of the chain break for the stabilo adjustment, the aircraft was caught in an uncontrollable flight attitude. Oblt Streiff was able to save himself with a parachute. Luckhause-Ottikon, March 3, 1940.

Deformation on the left wing of J-384. Fl Kp 15, Buochs, April 1941.

On April 4, 1941, during parking duty on the J-384, a deformation of both wings was determined. The explanations of the *Fl Kp* 15 revealed that this deformation had to result over a longer period of time. The aircraft was given over to the *Armeeflugpark*. The wings had to be newly covered.

In August 1943 the "regulations of flying with high performance aircraft" was decreed. It was noticed that at high speeds brusque control movements had to be avoided, because they could lead to overstraining of the airframe.

A tragic accident took place on July 7, 1944. A *Doppelpatrouille* of the *Fl Kp* 9 took off at approximately 0900 hours in Unterbach for attacking practice on the Kägiswil airfield.

During the attack of an anti-aircraft gun position *Oblt* Brenzikofer had to evade a Bücker Jungmann that was in a landing approach. The speed of the Me 109 amounted to approximately 600 km/hr, the altitude circa 400 m/G.

Through the brusque interception maneuver, the right wing spar broke. The wing tipped up, was torn off, and thereby knocked off the entire tail assembly. The J-334 still flew, repeatedly overturning, approximately 1 km further before the left wing also tore off, and the aircraft crashed. *Oblt* Franz Brenzikofer had no chance; he died in the wreckage of his aircraft. The remains of the J-334 were brought to Dübendorf, where an investigation took place.

A material fatigue was basically ruled out after only 187 cell hours. One assumed that, after the brusque control activity a forced rupture was provoked, in which the slotted wing loosened through the sudden enlargement of the angle of attack. Through this, a high ascending force abruptly developed that led to the break of the wing.

The pilots were once more urged to take heed of the "regulations of flying with high performance aircraft" from August 1943.

After a further incident, the *Kdt* of the *Fliegertruppe* took drastic measures: On April 3, 1945, after an interception maneuver of the J-379, a severe buckling of the covering in the area of the inner slotted wing mounting was found on both wings.

Oberstdivisionär Rihner demanded that periodically (every three months) the pilots draw attention to the regulations of flying with high performance aircraft. The matter was of such importance that even a personnel control kept an account of it.

The torn off left wing of J-334. Kägiswil, July 7, 1944.

The fuselage of J-346 on the Helling. During a hard landing on November 20, 1941, the fuselage was jolted, and had to be taken to Buochs for repairs.

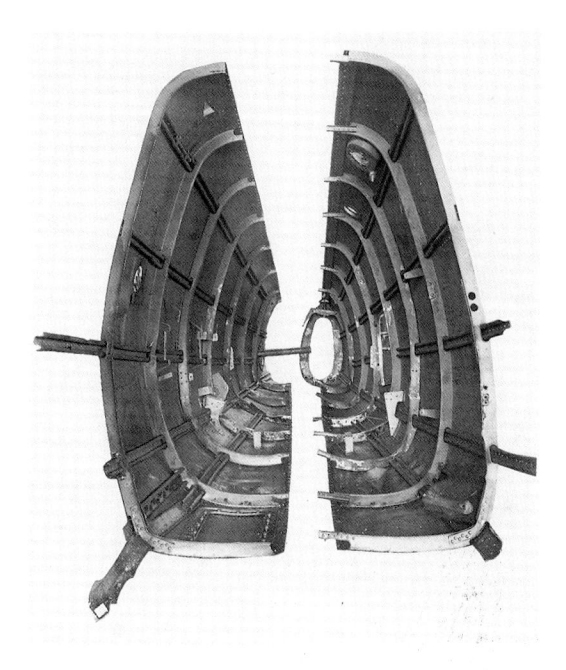

Stressed-skin fuselage of a Me 109 E. The design in two semi shells was a characteristic trait of the Me 108 and Me 109. To the lower left notice the access panel, normally hardly visible in photographs. To the far back, the continuous pipe for jacking up the aircraft.

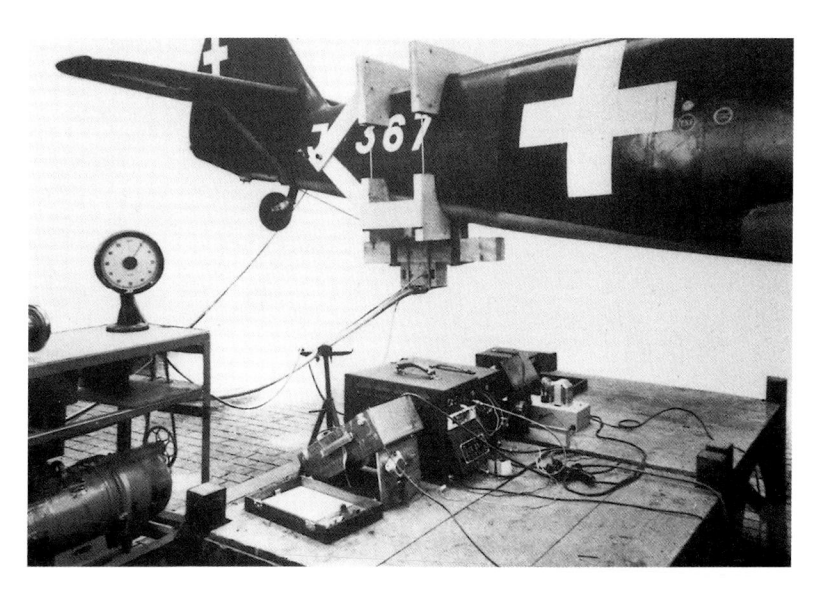

The fatigue test at Do-Flug in Altenrhein. J-367 suffered two accidents in 1944 with a large amount of material damage.

The protocol of an accident, Thun, May 1, 1940. During a hard landing due to a loss of speed, the landing gear on J-325 caved in. Lt Benoit, Cp av 6.

The fuselage was buckled in many places. J-325 was brought into the branch in Buochs for repairs. The repair work lasted until January 1942.

The aircraft was completely disassembled. Sections of the fuselage had to be newly constructed. The fuselage in the body shop on July 1, 1941.

The state of the repair work on November 10, 1941. At the height of the cabin, and on the rear section of the fuselage, numerous sheeting panels had to be replaced.

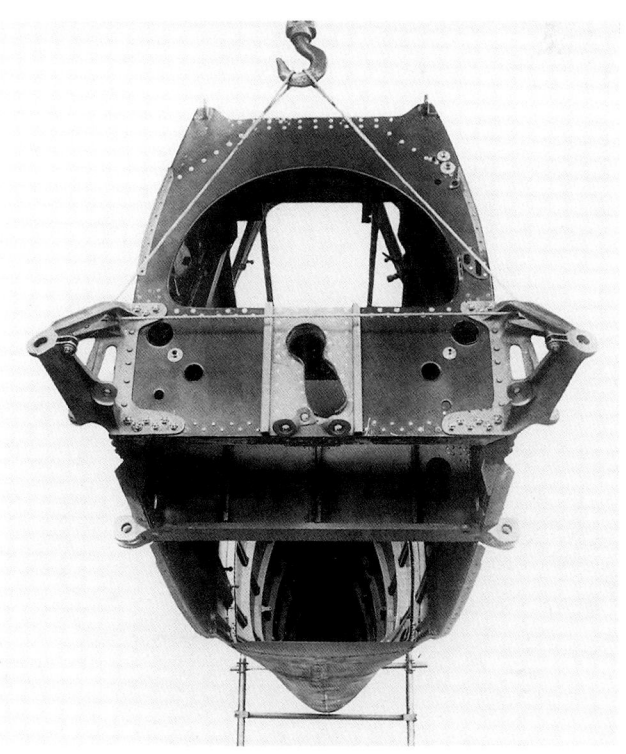

During the repair work, the wings and the motor were replaced at the same time. Furthermore, a S.F.R. construction was installed.

69

On December 23, 1941, J-325 was presented still without paint, but with the Kp emblem of the Cp av 6. One day later Adj Uof Urech slowed the aircraft down. 8,600 working hours were spent on the repair work. The amount of loss in francs is not known.

The completely revised J-325 in Buochs on January 19, 1942. The aircraft was flown in by Hptm Läderach on 27 January, and stated as ready for action on 23 February. In Emmen, on November 11, 1944, the landing gear of J-325 had broken, which caused damage of 20,000 francs.

The assembly hangar in 1941 in Buochs, the branch for the Me 109. The Armeeflugpark in Buochs was named "Gruppe Hug," after the then head of the site group.

The engine workshop in Buochs, with Daimler-Benz and Hispano engines. Notice in the foreground the U-shaped containers for the coolant.

A DB 601 on the engine test stand in Ennetmoos. Today, General Electric F-404 engines for the F/A-18 Hornet are also tested on the test stands with the most modern equipment.

The controls of the exhaust system on a DB 601. In summer 1941, instead of recoil nozzles, straight exhaust ports were experimented with.

It can be presumed this was caused from the development of acetic acid from ethyl acetate, as well as condensation water that developed during repeated warming and cooling down of the *K-Kraftstoff*.

To reduce the corrosion damage to a minimum, a range of measures came into effect: for the better distribution of the lubricant, the propeller had to be manually turned 3 1/3 rotations daily after the end of work. The engines were no longer turned off by shutting off the ignition, but rather through the closing of the fuel supply. However, this procedure was problematic, and did not apply for aircraft that were in on-call service. Furthermore, pulling the "quick stop" turned off the Daimler-Benz engines.

However, the *K-Kraftstoff* also caused damage to the lines. The Avioflex fuel lines were resistant to conventional gasoline, but were inelastic and brittle from the additions in the war gasoline.

At the beginning of 1944 the gasoline supply was precarious. From a total of 18,000 t aviation gasoline available, there were only 4,000 t OZ 87 and OZ 93, which could be utilized for the war aircraft. At a calculated daily consumption of 200 t, in the event of war this lasted for just 20 days.

In Summer 1944, the octane number K 93 was reduced to K 90 in the Me 109 E.

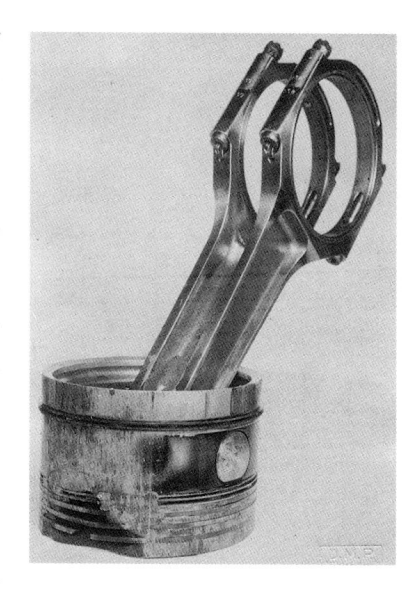

The use of gasoline OZ 87, with a lower lead content, led to heated or partially burned through pistons. As of January 1940, the aviation gas OZ 87 was used. Damaged piston from the engine DB 601 Aa No. 10903 (J-345) after only 3.5 operating hours.

The Jumo was furthermore operated with the customer aviation gasoline OZ 93, and did not change to K 90 until June 1945. Several months later, however, OZ 93 was again utilized.

According to the available documents, the Me 109 G was never operated with K gasoline. The *K-Kraftstoff* remained in use until Summer 1946.

As of mid 1946, the gasoline OZ 93 was no longer produced.

The Me 109 D/E were changed to OZ 100, while for the Me 109 G the remainders of the OZ 93 was utilized.

As of Spring 1943, the engines of the Me 109 E were converted to K fuel. Notice the inscription K 93 on the aircraft and barrel of gasoline. Due to the threat of corrosion, only dezincified barrels could be used. The Me 109 Es were fueled as of Summer 1944 with K 90, and as of mid-1946 with OZ 100.

74

Fl Kp 21 possessed a Cleveland "Cletrac" that was also utilized for gasoline transport. The wheels of the gasoline vehicle originated from liquidated DH-3 aircraft.

A camouflaged gasoline depot. Fl Abt 3, Avenches, Summer 1940.

The capacity of the gasoline barrel amounted to 250 l of fuel. The fuel tank of a Me 109 E contained 400 l. At a high performance, the DB 601 A consumed approximately 300 lt per hour. At a continuous output of 800 PS by 2200 U/min, it managed with approximately 240 l/h. Fl Kp 7, Avenches, Summer 1940.

75

The Installation of the Explosion-proof Oxygen System
The Me 109 D/E was equipped with a Dräger high-altitude breathing unit. Originally, two oxygen tanks with 2 liters contents were placed behind the flight deck. However, these light alloy high-pressure tanks could easily explode under an attack and do great damage.

This caused the KTA to develop an explosion-proof oxygen tank. The solution was found in a light steel round tank that was fixed on frame 4.

Although the tests were concluded by the end of 1942, it was not until 1944 that the first building alterations took place. The cost expenses for circa 80 Me 109s, 150 C 3603s, and 180 D 3800/01s amounted to a total of 420,000 francs.

Explosion-proof oxygen bottles made of light steel were available as of 1944. Because there are no pictures of a Me 109 available, a D 3800/01 is pictured in its place.

Pilot equipment in 1940. Notice the fastening points for the belts in the rear section of the cabin.

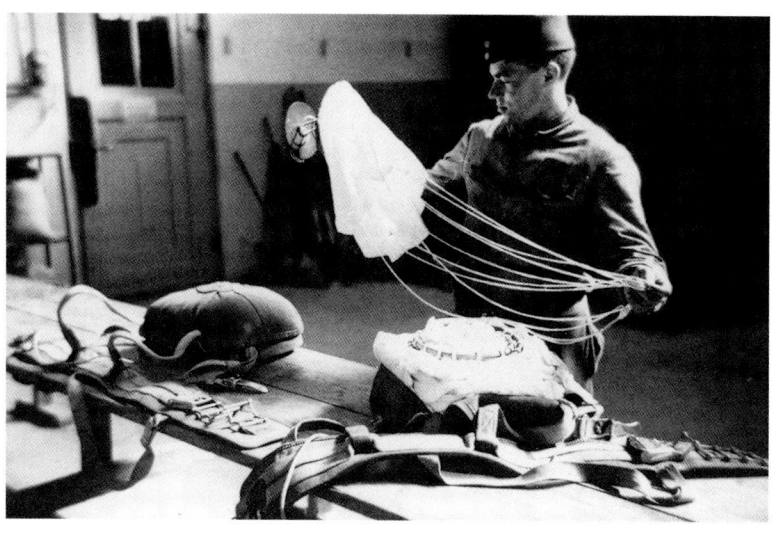

The first parachutes were introduced in Switzerland in 1928. None other than Hermann Göring, at this time as a representative for the Swedish company "Thörnbald," offered Switzerland a product. However, the Italian model "Salvator" was procured.

76

An Me 109 E as Jagdbomber in Dübendorf, June 1945. Notice the counter bearings of the 50 kg bombs.

The Emil becomes a *Jagdbomber*

Shortly after the beginning of war, the warring forces displayed the need to utilize fighter aircraft as bomb carriers.

In Autumn 1939, in Germany, the Me 109 V26 CA+NK was equipped with an ETC 500. Aside from the small ground clearance with a 500 kg bomb, the flight qualities also significantly worsened. The endeavor to utilize the Me 109 as a *Jagdbomber* led to the series with the abbreviation "B," or "Jabo," that could be loaded with a 500 kg, 250 kg, or four 50 kg bombs.

In the final phase of battle for England, the Me 109 E-4/B *Jabo* were intensively in operation.

The RAF followed a similar development. In North Africa they perfected the mission of *Jagdbomber* for close support of the troops. In the battle for Normandy in Summer 1944, the 2nd Tactical Air Force of the RAF caused true massacres with the Typhoon armed with bombs and rockets against the German motorized units. Just as on the Eastern Front the *Jabos* were not to be imagined without.

In Switzerland, this development was confronted for a long time. The *Flugwaffe* had numerous multi-purpose aircraft at their disposal that could carry bombs, but these models were already completely outdated at the start of war. A successor (the C 3603) did not reach the troops until Spring 1942.

In Spring 1941 it was discussed to use the D 3800 as a *Jabo*. However, three years passed until the Armeeflugpark had the first prototype aircraft with a bomb assembly at their disposal.

The relationship of the aircraft models to their purpose was unfavorable: 15 *Fliegereinheiten* were equipped with *Jagdeinsitzer*, and merely six of those with utility aircraft that could carry bombs.

After the Allied landing in Normandy in June 1944, the situation had to be newly assessed. Terrestrial invasions on Swiss territory were again in the arena of possibilities. The partially spectacular successes with the *Jagdbomber* on the various theaters of war most likely had an influence on the rethinking in Switzerland.

As of August 1944, all Me 109 Es were equipped with a bomb assembly. The reconstruction work took place through F+W Emmen and Pilatus Stans, and lasted until May 1945. Socalled "normalization" per aircraft lasted four to six weeks depending on the amount of additional revision work.

Two highly explosive bombs at 50 kg, or practice bombs made of Beton at 12 kg were used. Depending on the type of bomb, a corresponding counter bearing was installed. The detonation of bombs was accomplished with an instantaneous fuse, or a delay fuse with 0.3 seconds delay (appendix no. 14).

According to the bomb type, a corresponding counter bearing was installed. In the photograph a 12 kg concrete bomb identified with a green head.

Me 109 E Jagdbomber and D 3801 in Payerne in front of Hangar 1, Spring or Summer 1945. In the foreground a 50 kg bomb on a gimbal, next to it the detonator.

After the end of the war in Europe, the bomb assembly was only seldom used, particularly since the assembly was afflicted with technical problems.

In September 1945, all counter bearings for the 50 kg bombs were withdrawn and deposited in N+R Stans. However, several counter bearings for 12 kg training bombs were left on airfields that were authorized for shooting operations.

With the mission as a *Jagdbomber*, an old problem with the Me 109 once again appeared—the structural weaknesses. Brusque interception maneuvers during bomb release extremely strained the wings. Also, the propeller gusts of a pre-engaged aircraft could negatively affect the following aircraft if it was found in the interception phase of a bomb attack.

This led to the maximum permitted speed of the Me 109 E at 550 km/hr, and the forbiddance of bomb release in a dive. With these limitations, a mission as *Jabo* made little sense.

With the "Internal Communication No. 876" from February 6, 1947, all departments and site troops of the DMP were given instructions to disassemble the bomb equipment—only the release contact on the control handle and the label "BOMBES" were left.

Armor Protection in the Me 109 E

The Me 109 originally did not have armor protection—neither for the pilots, nor for the systems and fuel tank. Since the incident with *Oblt* Homberger in June 1940, however, armor protection in the Me 109 E was a lasting theme.

Also, in Germany the need for armor for protection of the pilot became apparent.

With the introduction of the E-4 model, the German pilots had for the first time (as of May 1940) armor protection at their disposal in the Me 109. Furthermore, this version had an enlarged canopy with a reinforced frame. In single aircraft bulletproof glass was additionally added on the windshield.

In Switzerland, the problem of armor protection was delayed due to internal rivalries and incompetence. After the fatal shooting of *Oblt* Treu on September 5, 1944, a feeling of insecurity became noticeable among the pilots. The concerns that they brought to paper on 12 October 12 received the response:

"The question of armor of the Me 109 was tested a long time ago. The additional application of armor is out of the question due to technical reasons."

However, in Germany in Summer 1940 the Me 109 E-3 was already subsequently provided with armor protection.

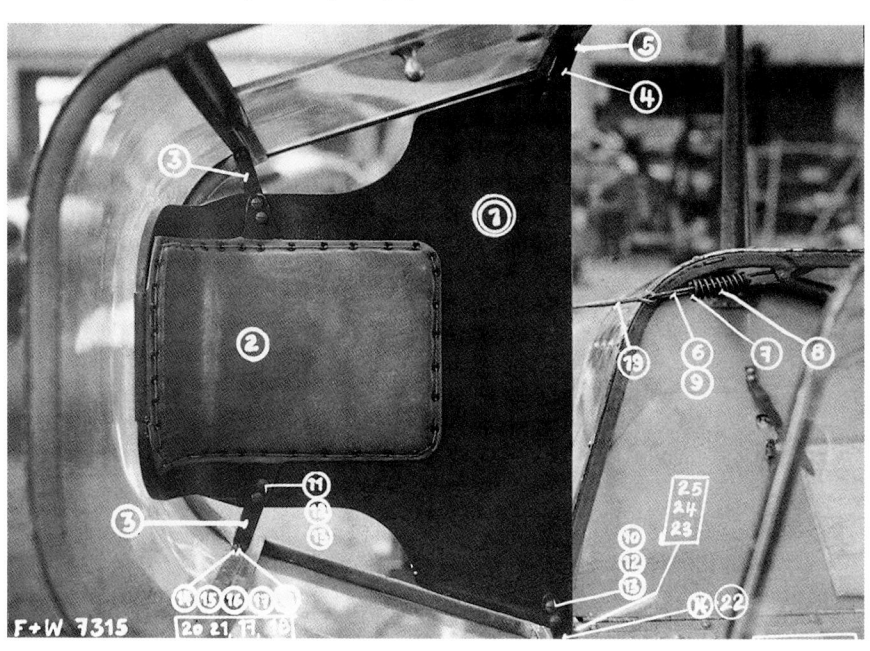

The vital armor plates were not installed until after the end of the war in Europe. F+W Emmen, May 1945.

78

The new 15 Watt UV lamp on the right side of the cabin (6). Night flight equipment was tentatively installed. F+W Emmen, July 1945.

The *Generalstabs-Offizier Hptm* Pista Hitz explicitly pointed out in a writing on October 23, 1944, the need for armor. *Oberstdivisionär* Rihner thereupon ordered the KTA to once again review the issue. They responded on 1 December: the installation of a head, shoulder, and back armor similar to that of the Me 109 F was estimated at approximately 1,000 francs per aircraft. The steelworks from Roll was to deliver monthly as of March 1945 circa 20 armor. An obvious problem was the additional weight of 35 kg.

Since the introduction of the Me 109 E, the weight increased by 90.6 kg, attributable to the installation of various devices. The multiple load had to be reduced from 6 to 5.8.

The installation of the armor took place after the end of the war. It is not known if all aircraft were re-equipped. Based on the weight, however, armor plates were again disassembled.

The Night Flight Equipment
The overflight of foreign aircraft at night proposed a special problem. For the time being, it was single reconnaissance aircraft or air freighters, and as of Autumn 1940 bombers of the RAF flew on the way to the target area in Italy through Swiss airspace.

Flying at dusk and in the night was certainly practiced, but the proper night fight equipment, as well as the corresponding ground organization, were missing. The radio measurement devices (the English term "RADAR" was then still not common) were not available. One managed with acoustic sound detecting devices and searchlights.

The tactics were similar to those of the German "light night-hunt" or "wilde sau." By means of searchlights from the ground, a carpet of light was laid out. Aircraft that flew over were seen from above as black silhouettes. The *Jäger* individually attacked the opponent like "wild sows."

The German *Luftwaffe* adopted these methods with variable success in Summer 1943. Switzerland could not effectively apply this tactic, however, because an extensive net of searchlights was missing.

In April 1943 the UeG began with the establishment of a *Nachtgeschwader*. The mission not only encompassed the night hunt, but also ground battle at night.

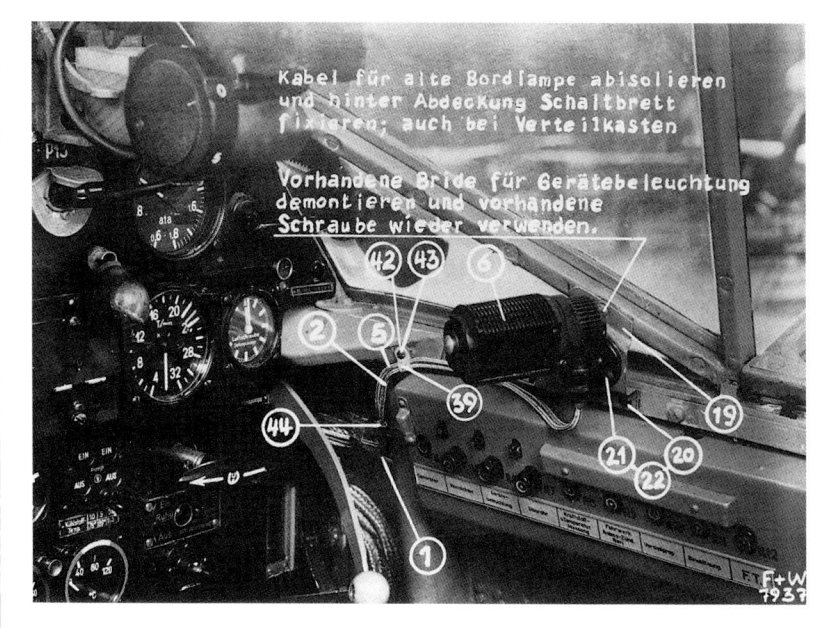

Night flight equipment: drop signal for the position lamps (15), cover for the fuel warning light (38), flip switch for UV lamp (8). New lamp above the compass, light border mark on the speedometer, shot counter with fluorescent color, crosshairs, and round front sight of the Revi 3c with fluorescent color. Notice the new gun selector switch and the switch to the signal rocket assembly "Fusées de Signalisation."

79

In 1944 J-391 was recreated from reserve parts and the fuselage of J-306. Notice the canopy with armor plate. Payerne, March 28, 1947.

J-393 in the assembly hangar at Do-Flug. Notice the wide blades of the EW propeller. Altenrhein, Summer 1945.

The J-306 received a replacement fuselage. The damaged fuselage was repaired in Buochs.

The J-391 was assembled from this fuselage and spare parts. Just three years later this aircraft crashed. Due to oversteering, the J-391 swerved at landing in Payerne on March 28, 1947, whereby the wings were slightly buckled. The damage amounted to 182,000 francs, and the aircraft was written off.

After the end of war in Europe, a stock of spare parts could moderately and peacefully be accounted for. Doflug, in Altenrhein, built eight further Me 109 Es from spare parts between July 1945 and March 1946. The aircraft were matriculated with J-392 to J-399, but did not receive any factory numbers. The aircraft largely corresponded to the original, but were equipped with an EW V6 propeller. Likewise, F+W Emmen built 17 D 3801s from spare parts between May 1947 and July 1948.

In January 1939 the AFLF was asked by the KTA if the construction equipment and gauges for the Messerschmitt aircraft could be liquidated. At this time the *Flieger-Staffel* 8, with 24 aircraft, was in action. A further manufacturing of component parts was out of the question.

The last reproduction Me 109 E was given to the Troop on March 13, 1946.

Chapter 10:

The Aircraft Procurement in the Forties

The Search for a Modern Fighter Aircraft

As was previously mentioned, in Spring 1940, due to the delays in the delivery of the Me 109 E, the procurement of fighter aircraft from Italy was considered. Pressure also came from the political side. Clearly, in Italy there were sensitive reactions, because up to this point Switzerland had not considered the Italian aircraft industry.

At the end of April 1940, 24-36 Macchi C 200 were offered to the *Flugwaffe* at a price of 260,000 francs each. Further offers took place in May for 30-50 Reggiane Re2000s and two dozen Caproni Ca 310/312s. During air combat between the *Armée de l'air* and the German *Luftwaffe*, the superiority of the Me 109 E in contrast to the MS 406 was displayed.

The improved MS 410 had a reinforced wing armament of four 7.5 mm MGs with ammunition belts in place of the barrel magazines, a firmly mounted cooler, and exhaust ports with recoil effect and further modifications. They arrived too late to still be used against the German *Luftwaffe*. Furthermore, the MS 410 had an inefficient HS 77 engine with 860 hp. The further developed MS 412, with the HS 51 with 1,000 hp performance, could no longer be built in France as a result of the events of war. This aircraft formed the basis of the D 3801, of which over 200 were built in Switzerland in Lizenz.

Although the first D 3801 were already complete in October 1940 there were problems, especially with the engine manufacturing. The delays extended to the year 1944. At this time the aircraft had become outdated by technical developments.

Also, the tried Me 109 E as of 1941 could no longer be considered as a fighter aircraft of the first and foremost. Aircraft like the Me 109 F or the Spitfire Mk V represented the standard. Because the Swiss aircraft industry could not develop fighter aircraft at short notice, suitable aviation material had to be sought abroad. The acquisition of complete aircraft, like the acceptance of a licensed manufacture, was considered.

In April 1941 a procurement of aircraft from the USA was again considered, whereby the following fighter aircraft were mentioned:

- Bell P-39 Airacobra
- Curtiss Hawk 81-A (P-40)
- North American NA-73 (P-51)
- Republic P-43 Lancer
- Vought XF4U-1 Corsair

Because the USA, at this time, was still not in a state of war with Italy, a transport of the aircraft via the sea route to Genoa was seen as possible. After the USA's entry into war in December 1941, however, this was no longer a possibility.

Furthermore, the Hawker Tornado, as well as an unspecified Spitfire, were mentioned. However, there was little hope of being able to procure aviation material from England. The same applied for future deliveries from Germany.

Controversial Licensed Manufacture for the German Reich

The newly founded aircraft factory Pilatus AG, in Stans, tested the possibility of a licensed manufacture of the Me 109 and Me 109 E and F for the German *Luftwaffe* at the end of 1940. The negotiations between Herr Bührle (owner of the engineering works Oerlikon, and 50 percent joint partner of Pilatus AG) and the Messerschmitt GmbH occurred at a time when the factory had still not taken up operations in Stans. The EMD fundamentally did not have anything against a licensed manufacture for their own requirements. Aircraft production for the German Reich, on the other hand, had to be refused for various reasons. The production of aircraft for the German *Luftwaffe*, as was debated, would also involve German personnel, including *Gestapo* agents, and this in the middle of an *Armeeflugpark*. The Elektrobank—25 percent joint partner of Pilatus, which also cultivated connections to Anglo-Saxon countries—likewise had no interest in aircraft production for Germany. On the other hand, it was debated that various companies in Switzerland were manufacturing for Germany, and the AG for Dornier aircraft in Altenrhein was completely under German control. The issue was obviously so politically significant that the *Bundesanwaltschaft* eavesdropped on certain telephone conversations. In a conversation on January 2, 1941, it could be gathered that a monthly number of items of ten to fifteen aircraft was scheduled. The price stated was 180,000 francs for a Me 109 F without armament, and 200,000 francs with armament. The conversation implied that business with Nazi Germany was not approved of everywhere.

The licensed manufacture was not achieved. Pilatus made a name for itself through its own products that today are significant worldwide. The Pilatus AG since then has also conducted repair and revision work for the Swiss *Luftwaffe*.

No good marks for the D 3801. Due to serious defects, it was classified in 1942 as unfit for war. Production of D 3801 in the K+W Thun in 1941. In the background a disassembled Me 109 E (J-383).

The Morane is Classified as Unfit for War

During a meeting with the head of the EMD, *Bundesrat* Kobelt, on January 7, 1942, *Oberstdivisionär* Bandi commented that the D 3801 was unfit for war!

This initiated a strong reaction from his archrival, the head of the KTA. Bandi's reproaches were, however, stark:

- At over 7,000 m/H the gasoline supply fails
- The MGs do not function at a greater altitude and in low temperatures
- The cannons do not function at deceleration
- The cooling system is poor
- Flaws with lubrication and bearings, thick oiling up of the spark plugs
- No radio equipment

What followed were denials, and again corresponding answers.

On 23 February Bandi had to call on the *Oberbefehlshaber* of the army, General Guisan. The reproaches applied to an aircraft of which 90 examples were delivered to the troops at this time. The older model, the D 3800, could no longer be seen as a predominant fighter aircraft.

With around 300 examples, the Morane remained, or at least outnumbered, the mainstay of the Swiss *Flugwaffe* until the introduction of the P-51D Mustang in 1948 and DH 100 Mk 6 Vampire in 1949. The last D 3801 were taken out of service in 1959.

During the year 1942, aircraft such as the P-47 B, Mustang Ia, Fw 190 A-4, and Spitfire IX were used in Europe. In the *Luftwaffe Helvetiens*, the Me 109 E-3 was still the most capable fighter aircraft.

The Macchi C 202

In March 1942 it was tested whether the Swedish aircraft industry produced a suitable product. It appeared that the Swedish *Luftwaffe* also had to procure their fighter aircraft from abroad, primarily from Italy.

By this time Italy produced efficient fighter aircraft with Daimler-Benz-Reihenmotoren. They aroused criticism only due to the weak armament.

At this time in the *Regia Aeronautica*, the Macchi C 202 was already used by the troops. This aircraft, powered by a DB 601 engine (built by Alfa Romeo in Lizenz), found increased interest from the Swiss procurement entity. Thereby, it involved an offer of 30 aircraft. In a document dated July 24, 1942, the *Kommandant* of the *Fliegertruppe* indicated that he deemed the price per item of 400,000 francs too high. Furthermore, he demanded that an inspector of the KTA monitor the production for a possible delivery.

In Spring 1943, the Swiss pilots could test the Macchi C 202 in Italy. However, a purchase was foregone. On March 1, 1944, a Macchi C 205 of the ANR (Aeronautica Nationale Repubblicana) landed on the Lausanne-Blécherette airfield. The aircraft was subsequently tested in Switzerland.

At the end of April-beginning of May 1943, a delegation under *Oberst* Högger could test the MC 202 in flight on Linate airfield, by Milan. The aircraft left behind a good impression regarding the flight performance, as well as the mechanical assembly. However, there were some doubts as to whether the MC 202 could be deployed on all Swiss airfields due to the relatively high area loading. The question also remained open if the armament of (two each 12.7 mm MGs and 7.7 mm MGs) was still opportune for a fighter aircraft.

For clarification of further details, the procurement of two aircraft at an item price of 370,000 was proposed. Furthermore, a possible delivery of 18 aircraft with corresponding spare parts was negotiated.

The Italian War Ministry immediately made demands for 8,500 liters of aviation gas. This brought the responsible parties into an unpleasant situation, On one hand, the Allies of Switzerland delivered important raw materials for the manufacturing of aviation gasoline, and even Italy's war opponents were supplied with these fuels. On the other hand, it also had to be accepted that both aircraft would not be delivered. *Divisionär* Bandi decisively opposed delivering the urgently needed aviation gasoline for the Swiss *Flugwaffe* to Italy.

The conclusion of this deal was made by a secret document from May 7, 1944, by the *Oberkommando* of the German *Luftwaffe*. The RLM in Berlin was asked what happened to the MM91992 and MM91993 aircraft, which were located on Linate airfield.

However, a Macchi arrived in Switzerland. On March 4, 1944, a MC 205 of the ANR landed in Lausanne-Blécherette. The MC 205 (MM92289) was flown to Switzerland on 18 flights circa 6.45 hours.

Negotiations with Germany

While a fighter aircraft remained in action on the Front for a relatively short time in the warring states, the aircraft of the Swiss *Flugwaffe* were increasingly made use of due to the longstanding operation. This attracted negative attention, above all with the Messerschmitt aircraft, because one was dependent on replacement parts from Germany. Due to the situation of the war, and also as political consequences of the incidents from 1940, the demanded replacement parts and spare parts were never delivered at the desired scale. In order to sustain airline operations, the manufacture of replacement parts had to be taken up in Switzerland. Because technical drawings were lacking, this was associated with disadvantages regarding terms and costs. The manufacture of certain component parts, such as chassis or engines, had to be completely left aside.

The lack of exchange engines unfavorably affected the operational readiness of the aircraft. During repairs or revisions of the engines the entire aircraft had to be decommissioned.

In October 1943, one tested the possibility whether the engines could be given to Germany to the manufacturer's works for revisions. Daimler-Benz made the suggestion to take over Swiss engines in exchange operations; for each engine delivered from Switzerland for revision, an already revised German engine would be delivered.

In March 1943 there was again contact with the Messerschmitt company. Once again, it concerned the delivery of spare parts that was still pending. Additionally, a delivery of 20 complete or disassembled Me 109 Es and 30 DB 601Aa engines was wished for. Furthermore, it was tested if a delivery of 20-30 Me 110 destroyers and a licensed manufacture of the Me 109 F or Me 109 G, as well as the DB 605 were possible.

In an investigation by the KTA in April 1943 of a procurement of 20 Me 109 Es and spare parts, the following was determined: the credit from July 1939 for the purchase of 50 Me 109 Es and spare parts of 22 million francs was obtained. For spare parts procurement, new agreements had to be negotiated, because this was partially annulled or never undersigned by Germany. Additionally, a price increase of 60 percent compared to 1939 was expected. Due to this inflation, the price of one aircraft was calculated at 455,000 francs if Germany delivered the airframes and engines, and the assembly took place in Switzerland.

If the airframes were manufactured in Switzerland, and only the engines were delivered from Germany, the price was estimated to be 350,000 francs per aircraft. These prices, as was explicitly noted, were to be observed only as guidelines.

Aside from the purely financial deliberation, the question of the dates was also decisive. The representative of the Messerschmitt company, Herr Brindlinger, also held out the prospect of the delivery of entire assembly groups from Germany. It remains to be seen if in Spring 1943 the construction of Me 109 Es was still to be accounted for.

As of February 1943, the Messerschmitt Me 109 G-6 was delivered to the German *Jägereinheiten*. These aircraft, driven by a DB 605, were the most built version of the Me 109, with over 12,000 exemples.

On 24 May *Div* Bandi immediately demanded the procurement of a Me 109 G with armament and radio equipment. Additionally, to each a 13 mm MG and a 20 mm cannon, ready to fire on a makeshift gun carriage, and the appertaining ammunition were added. Additionally, the Messerschmitt company was to promptly prepare the agreements for the delivery of 50 Me 109 Gs with spare parts and 30 spare engines. Furthermore, the licensed manufacture of Me 108 D, Me 109 G, and DB 605 were planned.

On the basis of the French Morane MS 450, the utility combat aircraft D 3802 was developed at Doflug in Altenrhein.

Divisionär Bandi is Discharged

The driving force for a further aircraft procurement from Germany was without doubt the *Kommandant* of the *Fliegertruppe*, *Divisionär* Bandi. However, for some time he was subject to massive criticism. As a previous artilleryman, he was never truly accepted among the flyers. In addition to this were the constant disputes with the KTA. The German offices also knew this. In a document "Secret *Kommando* Matters," from May 29, 1943, the German *Luftattaché* in Bern reported to its superiors a corresponding report.

The problem was that, above all, *Reichsmarschall* Göring prohibited the delivery of spare parts and aircraft into Switzerland. However, the *Luftattaché* brought attention to the fact that with this, one was not only undermining the position of the pro-German *Flieger-Chef* Bandi, but also running the risk that Germany could lose a certain influence over Switzerland.

The *Luftattaché* assessed the defense as absolute. This meant that Switzerland would stand upon their defense against the Allies, also. Highly developed German aircraft in the hands of the Swiss *Fliegertruppe* could thus be seen also as protection for the southern Reich's border.

The *Attaché* concluded the report with the words, "hereby it becomes recognizable that the German *Luftwaffe* now has either everything to lose or everything to gain in Switzerland...."

At the end of 1943 Hans Bandi was dismissed, and replaced by *Oberstdivisionär* Fritz Rihner. The *Fliegertruppen* for the first time had a *Kommandant* from their own ranks.

In 1943, the Swiss-German commercial transport faltered, because on one hand, it was lacking foreign currency, and on the other, Switzerland was also subjected to the pressure of the Allies. This affected, above all, the export of ball bearings and machine tools.

The endeavor to procure further aircraft and spare parts from Germany was carried forward. In October 1943 two brand new Me 108s, and two used but revised Me 109 Es, were offered to Switzerland.

The price for a new Me 108 amounted to 79,000 Reichsmark. In 1939 a Me 108 cost only 31,000 Reichsmark. For an Occasion Me 109 E 168,000 was demanded. One attempted to bring down the prices, but did not further enter into a deal.

One intended to exchange both Me 109 Fs that landed in Belp in 1942 with two Me 109 E-7s from Germany. Later, a purchase of both Me 109 Fs at approximately 300,000 Reichsmark was considered, along with a delivery of corresponding spare parts. Both were never achieved.

After long and tough negotiations, the long awaited spare parts for the Me 109 E arrived, at least partially.

Instead of the demanded 20 new engines, only five revised Occasion engines at an item price of 42,000 were delivered. In addition, crankcases, device carriers, cylinder blocks, and items not manufacturable in Switzerland at a value of 237,840 Reichsmarks were delivered. The urgently needed crankshafts were, however, furthermore not released for export.

No Chance for Proprietary Development

At the end of 1943, the operational readiness of the Swiss *Flugwaffe* was not satisfactory. All war aircraft in operation were afflicted with some kind of flaw.

From the still circa 70 Me 109 Es, no more than 28 were completely ready for operation. The remaining aircraft, respectively their engines, were either undergoing repairs, or had no radio equipment installed.

The better part of the *Flugwaffe* consisted of the Morane D 3800/01 (MS 406/412) fighter aircraft. Based on the Morane-Saulnier MS 540, Doflug in Altenrhein developed a new utility combat aircraft in 1944.

The MS 450 was the last stage of development of the type series MS 405 before the war. The French air force, however, preferred the modern Dewoitine D 520.

The Hawker Sea Fury Mk X in Dübendorf, July 1946. The Bristol Centaurus 18 engine, with its over 2400 hp, satisfied. However, the time for propeller driven fighter aircraft drew to a close.

Already in October 1939, the KTA planned a potential delivery of 100 aircraft from France. The delivery of the 2nd series Me 109 E, however, rendered this plan superfluous.

With the MS 450, which was further developed in Switzerland, a Saurer aircraft engine YS 2 with 1250 hp came to the assembly. The aircraft, named D 3802, took off on September 29, 1944, for its first flight. Technical problems, especially with the YS 2, however, delayed the commissioning. An O-series of eleven aircraft came to the troop as of 1947 under the name D 3802A. The D 3803 remained the single item with the newly developed Saurer YS 3 engine. Although this aircraft was at least satisfactory regarding performance, the technical problems would never be eliminated.

The development of aircraft with jet propulsion also did not remain hidden in Switzerland. Already at the end of the 30s it was tested whether the gas turbines would be utilized as engines in the aircraft. However, there was still no time for the development of their own aircraft with jet propulsion.

In Germany and England, the aircraft had been in troop service since 1944. Projects of Swiss provenance had no chance. The blueprints of the Doflug/FFA P-25 and P-26 had just as little chance as the N-10 and N-11 of F+W Emmen.

In Summer 1946 Switzerland could procure from England a De Havilland DH-100 Mk I. During an accident at takeoff on August 2, 1946, the aircraft was broken. In Spring 1947, however, another DH 100 was available, and Troop testing could be carried on. Subsequently, as of 1949 over 175 De Havilland DH 100 Mk 6 "Vampire" were procured.

Aircraft with jet propulsion, or short "*Düsenjäger*," were at this time certainly advanced. Regarding the performance, aside from maximum speed and climbing capacity, however, they were in no way superior to the piston engines of the last generation. In addition to this was the faulty reliability of the engines, and the enormous fuel consumption. Furthermore, in Switzerland the problem with the mountain airfields and the narrow mountain valleys presented itself. From these considerations, one continued to be concerned also with the propeller drive.

In 1946 in Langley (UK), the Hawker Fury was tested. In July of the same year a Sea Fury Mk X was demonstrated in Switzerland.

When at the end of 1947 the opportunity was offered to buy a P-51D Mustang from the stocks of the USAF, for the last time a fighter aircraft with a piston engine was procured. From 1948 to 1957 130 Mustangs were put into operation without any noteworthy problems.

At the end of the 50s, the last aircraft with piston engines from the *Fliegerstaffeln* were taken out of service.

In 1947 Flieger-Staffel 17 was equipped with D 3802/03. Although the aircraft were afflicted with technical problems, they remained in service in the troop until 1956. In the photograph is an individual D 3803.

Two MS 406 were delivered at this point of the mobilization. They were at the disposal of the K+W Thun, and were not with the troop.

A fortunate circumstance was that the *Fl Rgt* 2 was attending a refresher course on its bases in the areas of Bern and Western Switzerland, and could immediately secure the border control. The Me 109 Jumo of the *Fl Kp* 15 in Payerne were the single modern aircraft of the entire force.

Each with six Me 109 Es, on 29 August the Cp av 6 mobilized in Thun, and the Fl Kp 21 in Dübendorf.

On 30 August the General was elected and sworn in by the united Federal Assembly. During this ceremony, five Me 109s of the Fl Kp 15 from Payerne were circling above the *Bundeshaus* in Bern.

With the general mobilization on 2 September, the remaining units—also without aircraft—moved into active duty.

The pilots and look-outs of the *Flieger Kp* without aircraft were split up to other units for further training, while the ground crew, after approximately three weeks, was again dismissed.

The three units equipped with Me 109s were immediately deployed to border control.

The *Fl Kp* 6, 15, and 21 *Alarmpatrouillen* were ready, in flying weather, for the control and blocking of Swiss airspace.

The locations of the three Messerschmitt units, as opposed to the field airports, were provided with a good infrastructure. Dübendorf was the center of military aviation since the end of 1914, and had a concrete runway available. Since 1938 Payerne was the location of the pilot recruiting school, and in Thun the Swiss construction workshop found its home.

In order to prevent violations of the border by the Swiss military aircraft, the restricted areas were set up along the border. In these areas the *Jagdpatrouillen* were permitted to fly only with the authorization of the *Armeekommando*.

Until the end of 1939, 143 violations of the Swiss airspace were reported. The Me 109 *Patrouillen* could not prevent this, because the overflights occurred either during the night, in bad weather, or in the restricted border zones.

Until the end of 1939 approximately 1450 hours were flown in over 6,000 flights on the Messerschmitt aircraft.

Despite intensive flight operations, there had been no severe accidents recorded since the mobilization. On the other hand, the familiar engine problems were present.

The Messerschmitt units at mobilization 1939:

Rég d'av 1	the	CP av 6 with	6 Me 109 E in Thun
Fl Rgt 2	the	Fl Kp 15 with	6 Me 109 D in Payerne
Fl Rgt 3	the	Fl Kp 21 with	6 Me 109 E in Dübendorf

Emergency landing due to a defect in the propeller adjustment. J-333 had only 6.33 operating hours, and was still not equipped with weapons. Zimikon, August 31, 1939,

90

J-325 and J-326 in the first winter of war 1939/40.

Deceptive Peace

The year 1940 was probably the most turbulent in the history of the Swiss *Fliegertruppe*. It was the first trial by fire that the *Flugwaffe* had to pass.

In the first two months air service was considerably affected by bad weather. The airfield were partially unusable due to frozen, softened or flooded runways. The number of airspace violations was comparatively minor with ten reports in January and nine in February.

In order to control the effectiveness of all the *Einheiten* of the F1.B.M.D and *Fliegerverbände* deployed for neutrality protection, on 13 January extensive resistance practice was carried out.

Across Switzerland—from the stretch Lake Constance - Genfersee and back—a Junkers Ju 52 simulated border violations. Fourteen C 35s and eight Me 109 *Patrouillen* were thereby deployed. The communication network C was overloaded with over 15,000 reports. Four Me 109s and six C 35 *Patrouillen* could find the Ju 52.

To make the nationality of the Swiss aircraft more easily recognizable from the ground, as of mid-January the enlargement of the national emblem on the wing's underside was applied, and additionally the entire rudder was painted red.

The Fl Kp 7, 8, and 9, until then without aircraft, were newly organized. In February, in Geneva, under *Hptm* Wilhelm Frei, the retraining on the Me 109 began. During air operations there were repeated violations of French airspace, which as a consequence yielded complaints from France.

After retraining, six units with eight to twelve Me 109s each were ready for action. Thanks to the better weather in March, air service could begin again normally on all airfields.

In March, the Swiss airspace was violated in 19 cases. One could determine that the air defense was no longer optimally organized. A German He 111 could fly over the Dübendorf military airfield in the protection of the clouds without being prematurely discovered. The *Alarmpatrouillen*, which ascended too late for defense, could no longer intercept the intruder.

Another resistance practice was scheduled for 2 March. With a Potez 63 as the target aircraft, six C 35s and a Me 109 *Patrouille* could intercept the Potez. Also, this exercise displayed that the means of communication were not sufficient for the mission.

At the beginning of April, the order was decreed to attack military aircraft of the leading warring states without forewarning. Aircraft of non-leading warring states continued to receive warnings, in which an aircraft of the *Abwehrpatrouille* would overtake the intruder from the side and, with wobbling of the longitudinal axis attract attention. This risky maneuver was imposing, because the Swiss Me 109 at this time did not have any equipment for shooting signal rockets at their disposal.

At the beginning of the year 1940 the air traffic was strongly limited because of bad weather. Fl Kp 21, Dübendorf.

91

Fl Kp 15 was the sole unit that was delivered the Me 109 D as fighter aircraft. The 15s remained stationed in Payerne until the end of March 1940.

The retraining took place under the leadership of Hptm Wilhelm Frei (later known as "Düsen-Willi"). Geneva, March 11-16, 1940.

The pilots of the newly organized Fl Kp 7, 8, and 9 retrained on the Me 109 in March 1940. During airline operations in Geneva, it repeatedly came to violations of the border with France.

Retraining aircraft Me 108 Taifun in Geneva. Fifteen Me 108 were procured in Germany. In the course of the war three further were added by purchase and internment.

In Spring 1940, Fl Kp 21 was stationed in Dübendorf and placed Alarmpatrouille in Olten and Mollis. Mid-May they transferred to Emmen, but kept aircraft at the Dübendorf airfield on alert.

A Me 109 E of the Alarmpatrouille of Fl Kp 21 in Olten, April 1940. In the background a EKW C 35 of Fl Kp 16.

For better identification, in April 1940 the aircraft were newly marked. On the fuselage side a Swiss Cross was added, and the number was minimized by 25 cm.

93

For a propaganda film of the Fl.B.M.D. in 1940, it was effective for the Me 109 E of Cp av 6 to intercept a Poetz 63. The takeoff preparations of the Potez 63 in Thun.

Me 109 E of Cp av 6 in an attack. The sequence originated from a 16 mm film. A C 35 served as a camera aircraft.

The Potez was equipped with a smoke system. This consisted of a pipe with a funnel, from where soot was sprayed.

However, during a downturn maneuver soot came quite suddenly into the cabin, caused by the dynamic pressure. The cleaning of the Potez lasted days....

Oblt Boudry (pilot) right, and André Kehrli, the flight mechanic.

The identification of Swiss aircraft was obviously a problem, because the *Fliegertruppe* flew the same aircraft models as the French *Armée de l'Air* and the German *Luftwaffe*. In order to increase the recognizability, as of mid-April the Swiss cross was applied to the fuselage side of all military aircraft, and at the same time the identification number was minimized to 25 cm.

In April, two Allied and fourteen Axis Power aircraft invaded Swiss airspace. Twenty-five further aircraft could not be identified. The *Alarmpatrouillen* did not succeed in catching the intruders, because the border violations mostly occurred in the still restricted border zone for the Swiss flyers.

The First Shots on Intruders

On the Friday before Pentecost, May 10, 1940, at 0535 hours the German offensive against the West began.

For this deployment *Luftflotte* 2 under General Kesselring, and *Luftflotte* 3 under General Sperrle had over 3,000 aircraft at their disposal. Among these were approximately 1,000 Messerschmitt Me 109 E-1s, E-3s, and the newly introduced E-4s.

The air combat again displayed the superiority of the Me 109 E. The Morane MS 406, which was praised by the *Armée de l'Air* as "meilleur chasseur du monde," did not meet expectations.

At the proclaimed second general mobilization, the *Fliegertruppen* were already in service. Through a regrouping, various *Fliegereinheiten* were transferred from the airfields near the border into the interior of the country.

The readiness for war of the Swiss *Fliegertruppe* was seen as en bloc, and also not optimal at the second mobilization. Only eight units were equipped with approximately 70 more or fewer modern aircraft, not all of which were available. This stood in stark contrast to the high war morale of the pilots and of the ground crew.

The will to defend the homeland against each aggressor made an impact when the forbidden area that hindered the mission of the *Alarmpatrouille* was lifted with immediate effect.

During the early morning, at 0635 hours on 10 May, Lt Thurnheer from Cp av 6 in the Brugg-Basel area attacked a German aircraft whose model he could not identify. If the aircraft (a Ju 88 of KG 51?) was hit was unable to be determined—it escaped over the country's border. Thus, from a historical perspective, Lt Thurnheer was the first Swiss pilot who shot at a foreign aircraft.

The first shooting down of an intruder took place on that same evening. A He 111 of the III./KG 51 attempted to fly over Switzerland in a brisk manner from the Ajoie to Lake Constance. *Hptm* Hörning and *Oblt* Ahl from *Fl Kp* 21 intercepted the aircraft by Bütschwil, and fired several warning shots.

The positions of the Jägereinheiten as of 11 May were as follows:

Cp av 3	D.27	in Belp
Cp av 4	D 3800	in Thun
Cp av 5	D 3800	in Thun
Cp av 6	Me 109 E	in Thun
Fl Kp 7	Me 109 E	in Payerne
Fl Kp 8	Me 109 E	in Avenches
Fl Kp 9	Me 109 E	in Avenches
Fl Kp 12	D.27	in Grenchen
Fl Kp 13	D 3800	in Geneva for retraining, not ready for action
Fl Kp 14	D 3800	in Geneva for retraining, not ready for action
Fl Kp 15	Me 109 D/E	in Biel-Bözing, as of the end of May in Olten
Fl Kp 18	D.27	in Luzern
Fl Kp 19	D 3800	in Geneva for retraining, not ready for action
Fl Kp 20	D 3800	in Geneva for retraining, not ready for action
Fl Kp 21	Me 109 E	in Emmen, Alarmpatrouillen in Dübendorf, Olten and Mollis

Due to the increasing air space violations, the *Fliegereinheiten* were placed on high alert as of mid-May 1940.

The Germans attempted to escape with aviation tricks in a low level flight. The aimed attack of both Me 109s made an impact, and the He 111 went to the ground with a strong trail of smoke towards Altenrhein, in German (Austrian) territory.

On this day, III./KG 51 had the assignment to attack French airfields in the Dijon and Dôle regions. The 9K+DR of the 7. *Staffel* under *Oblt* Schifferings lost their way on the flight back into Switzerland, and believed they had German Me 109s in front of them. The consequence of this error was approximately 9 cannon and 50 MG hits of both Swiss Me 109s, as well as a severely injured crew member. Of serious consequence was the error of a chain of 8. *Staffel* of KG 51. Due to poor visibility above the Black Forest, they deviated from the flight path and mistook Freiburg in Breisgau for Dijon. Instead of on French enemies, the bombs fell on their fellow countrymen. The German propaganda, however, was capable of attributing the attack to the Allies; a fact that was not completely clarified until just after the war.

The inhabitants of Courrendlin escaped in horror as a German He 111 mistakenly released 18 bombs early in the morning at 0520 hours. Only property damage arose on the overhead contact line of the Moutier-Delémont railroad.

Due to the increasing airspace violations, as of 16 May the *Fliegereinheiten* were on high guard.

This required that all aircraft ready for flight could be started within minutes one hour before sunrise until one hour after sunset.

The spectacular successes of the German airborne forces in Belgium and Holland demanded defensive actions on the airfields. The infantry training was disliked among the troop.

On 16 May another highlight took place: For the first time an aircraft was shot down that went down on Swiss territory. A He 111P of the 9./KG 27 was caught in a snowstorm in the Epinal region after an attack. During the attempt to reach home base in Baltringen—in the South German area—in a blind flight, the pilot lost his way in Switzerland.

On the early morning of 10 May, Lt Thurnheer of Cp av 6 was the first Swiss pilot to attack a foreign aircraft. In the photograph is Hans Thurnheer in front of a Fokker CV.

After a report of the Fl.B.M.D. the *Alarmpatrouille* of *Fl Kp 21* in Olten took off at 1715 hours, but could not find the He 111 due to the bad weather.

More successful was the *Alarmpatrouille* from Dübendorf. *Oblt* Streiff and *Oblt* Kisling discovered the aircraft at the upper end of the Greifensee, and were positioned circa 500 meters behind the He 111 in a higher attacking position. When the German aerial gunner opened fire, Streiff descended with the J-349 for the attack. Badly shot, and with injured crew members, the Germans were able to escape into the clouds. However, the bomber was caught in the 20 mm resistance fire of Flab Bttr 34 in the Dübendorf region, whereby he was hit repeatedly. The radio operator, as well as the flight mechanic, jumped with parachutes in the Ottikon region, and had to be brought into the *Kantonsspital* Winterthur with minor injuries. The aircraft crashed at approximately 1730 hours by Kemleten, whereby the left wing was torn off.

Hptm Hörning (standing right) and Oblt Ahl (standing left) of Fl Kp 21 shot down a foreign aircraft for the first time on the evening of 10 May.

He 111 P (1G+HT) of 9./KG 27, which made an emergency landing at Kemleten. The German occupying troops succeeded in setting the aircraft on fire.

The emergency landing initiated a general bustle.

The pilot and the lookout destroyed the most important flight instruments, and set the aircraft on fire. After a four hour getaway, a *Patrouille* of *Dragonerabteilung* 12 picked them up.

Both Me 109s landed in Dübendorf during a light blizzard. At the control of the aircraft *Oblt* Streiff determined a high in the propeller hub on J-349.

The remains of the He 111 attracted great interest. Approximately 200 hits were assessed.

However, the projectile effect of the 20 mm FF-K was not seen as convincing. The armor of the He 111 was sufficient in hindering damage to the vital parts.

As pilots are, they wanted to secure a trophy, and attempted to disassemble various parts from the He 111. Also, a national emblem from a wing was cut out. Thereby, tensions arose with the rescue team and the guard detail. *Hptm* Hörning could not even utilize the power of his rank, and the undertaking was discontinued.

The pilots of *Fl Kp* 21 had the opportunity to speak with the German crew. Pilot Lt Riecker, a veteran of the Poland campaign, stated "In Poland I was shot down three times, and each time I succeeded in returning back to Germany—and here, in tiny Switzerland of all places, this had to happen to me, damn it!"

The injured aerial gunner, *Oberfeldwebel* Hobbie, cursed, "if the pilot had flown the damned thing better, I would have already burned one of the little Swiss."

Thereby, it clearly emerged that German aerial gunners shot at each aircraft that took an attack position. The radio operator Herzig underlined the German position clearly "There is absolutely no landing order! Either the effects of enemy weapons force us, or we succeed in a bunk."

Both Me 109 fired circa 1200 GP 11 and circa 120 shot 20 mm shells. At the evaluation of the hits, however, the performance of the FF-K was not convincing.

At the end of May *Fliegereinheiten* received further reinforcement. The *Fl Kp* 13 and *Fl Kp* 14 could be operationally deployed after retraining on the D 3800.

However, the *Alarmpatrouillen* were still supplied by the experienced *Messerschmitt-Kompanien* 6, 15, and 21.

Problems with the Border in Jura

On 1 June KG 53 "Legion Condor" had the assignment to bomb the railway connections in Rives by Grenoble (I. *Gruppe*) and Ambérieuen-Bugey (III. *Gruppe*).

Item 7 of the mission order read, "At the targets only weak defense is to be reckoned with. Caution when flying over Swiss territory. Attack by Swiss fighter aircraft Bf 109 is to be reckoned with."

The German operations planners apparently did not seriously calculate the determination of the Swiss anti-aircraft defense: sections of the *Geschwader* flew in defiance of the Swiss air sovereignty at 1548 hours in the Basel region over the border, and left the country by Le Brassus.

With this maneuver, one wanted to obviously evade the French anti-aircraft defense, and at the same time shorten the way to the target regions.

Hptm Roubaty, *Kommandant* of Cp av 6 in Thun, took the initiative after a report from the Fl.B.M.D., and at 1605 hours took off with Lt Wachter in the direction of Neuchâtel. *Hptm* Roubaty's aircraft was equipped with radio equipment, while Lt Wachter had to communicate by hand signals.

At 1621 hours Wachter sighted a single He 111 at circa 3,000 m in the Vue des Alpes region. With a signal he informed the *Patrouillenführer*, who immediately led into an attack. Roubaty easily approached the bomber left from the rear, followed by Wachter, who attacked directly from behind. The German aerial gunner likewise opened fire, without causing an impact, however. For the time being, the He 111 was able to escape through the clouds in a nose dive. During the second attack of the Me 109, still circa 100 m above the ground, more hits could again be made. In a low level flight, the bomber got caught in a telephone line, and after several hundred meters crashed into a somewhat higher situated forest by Lignières. The He 111 exploded, and the debris tore a forest aisle over 200 meters.

Thereby, the five man crew was killed. The He 111 H with the marking A1+DM belonged to the II. *Gruppe* of KG 53.

Why the aircraft flew alone in a Northern direction keeps the question open if the aircraft had technical problems, or was damaged over France. The *Groupe de Chasse* II/7 that first was in action with Dewoitine D.520 reported on this day five shoot downs of He 111s, and the attack on another aircraft that was later shot down in Switzerland.

Oblt Viktor Streiff, Fl Kp 21, was engaged at Kemleten during the shooting of the He 111. It was the first shooting of a foreign aircraft that fell to the ground in Switzerland.

The Heinkel He 111 H (A1+DM) of II./KG 53 shot down by Cp av 6 at Lingniéres, June 1, 1940.

The motto of Cp av 6 "sans soucis" obviously corresponded to this hectic period. Oblt Hadorn on alert, Thun, Spring 1940.

During the flight back from their attack targets in France, KG 53 flew again along the Swiss border in the direction of Pruntrut-Basel.

1. *Staffel* under *Hptm* Allmendinger flew at 1708 hours in the region of Lac de Joux, in Swiss airspace.

Five minutes later, the message came in from *Flieger Kp* 15 in Olten concerning a unit of 12 German bombers. The 1. *Alarmpatrouille*, with *Oblt* Kuhn and *Lt* Aschwanden, took off immediately. Minutes later another German unit was reported.

The 2. *Alarmpatrouille*, with *Hptm* Lindecker and *Oblt* Homberger, took off at 1718 hours, and were ordered to Saignelégier.

Approximately at the same time, Lts Thurnheer and Schenk of Cp av 6 took off from Thun.

Patrouille Hptm Lindecker/*Oblt* Homberger discovered in the St. Imier region at circa 4,000 m altitude a formation of 4 x 3 bombers flying along the border. From the French side heavy anti-aircraft fire could be determined. Because *Hptm* Lindecker wanted to avoid a border violation on his side, he positioned himself behind the unit without attacking, and waited until they arrived at Doubs-Bogen by St.-Ursanne. Because they were unmistakably over Swiss territory, Lindecker attacked the rearmost bomber of the left chain, and was able to make several hits.

Lt Schenk, from Cp av 6 in the St.-Imer region, lost his *Patrouillenführer*, who disappeared in a cloud. Unexpectedly he saw the bomber directly approaching him. Schenk pulled his aircraft high, positioned himself behind the unit, and initiated an attack on the rearmost triple formation. Thereby, he was caught in the resistance fire of the aerial gunners, and his aircraft was repeatedly hit.

Lt Schenk concentrated on the bomber flying as the rear aircraft. The bomber suddenly turned off to the right, and proceeded into a steep gliding flight with a gray trail of smoke. The He 111 crashed by Oltingque, on the French side.

On 1 June, aircraft of Fl Kp 15 above Les Rangier stood for the first time in defense against German aggressors.

Pilots of Fl Kp 15 at the Olten base.
Oblt Suter, Oblt Rufer, Hptm Lindecker (Kdt Fl Kp 15), Lt Aschwanden, Lt Breitenmoser (Tech Of.), and Hptm Lindecker shot down three German aircraft above the Jura during combat operations.

Hptm Lindecker attacked the flying bomber at the right rear, and fired the remainder of his cannon munitions from approximately 300-200 m without, however, determining an impact.

Because the Swiss territory was flown through, the attacks had to cease. It was exactly 1743 hours.

The Swiss, like the Germans, believed to be in the right regarding the border violations. The attack on the bombers clearly took place above Les Rangier, which was confirmed by observers on the ground. Wachtmeister Hotz, from *Infanterie Kp* I/232, was nearly hit by a shot that exploded on the ground at his observation position in Bourrignon. His statements, as well as those of *Zugführer* Lt Favre, were put on record.

Hptm Lindecker later noticed in his combat report, "during the flight I noticed new French anti-aircraft fire, and suddenly also two Me 109s, that in the middle of this fire lunged at the *Staffel* without success."

At the same time, Swiss *Jagdpatrouillen* and French anti-aircraft combated the German bombers.

The already mentioned Wm Hotz reported that behind his back a shot exploded while he observed the combat in the air.

The investigation of the fragments revealed that it was no bullet from Swiss production.

The Heinkel He 111 was at this time equipped with the 7.92 mm MG 15, which did not fire explosive projectiles. But Hotz spoke of an explosion. This could mean that the French anti-aircraft also shot over the Swiss border. Were the border violations intended or not?

The line of the border in the Neuchâchtel Jura was defined by the Doubs. At Pruntruterzipfel, in Clos du Doubs, the river flows into Switzerland, goes into a wide bend to St-Ursanne, and leaves the country at Brémoncourt. The narrowest location of the Pruntrut measures approximately 15 km—even a *Bomberverband* could cover the stretch in circa 3 minutes. From the Doubs bend by St-Ursanne until the border in the northeast it was 8 km, and from Les Rangier 5 km.

Soldiers of Fl Kp 15 during construction work on the Olten base.

100

Uffz Mahnert's He 111 P-2. The right engine was damaged over France and had to be set aside. Ursins, June 2, 1940.

The French flak was very strongly concentrated in the Belfort area. In order to avoid this danger, KG 53 flew closely to the Swiss border. A *Kampfgeschwader* in an unwavering unit flight, however, had kilometer wide dimensions, and a change in course over the Pruntrut was purely impossible from a tactical perspective. Thus, single aircraft, deliberately or not, came over Swiss sovereign territory.

Germany had at their disposal a widespread net of radio beacons for navigational purposes. These undirected medium wave transmitters gave off a continuous radio signal. The navigators needed only their radio direction finder (i.e. PG 6) set to one or several transmitters, either to report their own position, or to determine the flight path.

One of these radio beacons was located by Freiburg, in Breisgau. When one drew direct lines to the targets of attack of KG 53, they followed exactly the Jura range, and crossed the Pruntrut.

The approach and the flight back of the German bombers in Swiss airspace took place practically at the same position. This led to the assumption that the navigators utilized this "electronic lighthouse" as an orientation guidance, and simply ignored Swiss air supremacy.

June 2, 1940 - The Odyssey of Uffz. Mahnert
On this Sunday morning, between 0830 and 1100 hours, numerous reports of border violations emerged. *Fl Kp* 13 and 14, equipped with D 3800, were additionally put on high alert. None of the aircraft, however, had radio equipment at their disposal.

At first no closely identified bomber flew coming from Geneva at 7,500 m altitude over the entire midland, and could not be placed by any *Patrouille*. The mission of *Fl Kp* 13 as *Sperrflieger* in the Biel-Lyss region proceeded inconclusively, because the D 3800 could not reach the necessary altitude in sufficient time.

The *Kampfgeschwader* 55 "Greif" in these days was deployed for the prevention of military transports from southern France, and for the destruction of airfields in the regions from Paris to Lyon. III./KG 55 thereby had the assignment to attack the Bron airfield by Lyon. Over Bourg, the He 111 P-2, flown by *Uffz* Mahner, was attacked by a Dewoitine D.520 of GC II/7, whereby the radio operator was injured. Furthermore, the right engine was hit, and had to be taken out of service due to oil loss. Mahner tried to continue to fly with the left engine that after some time overheated. Due to the loss of power, the aircraft constantly lose in altitude. An overflight of the Alps into allied Italy was not possible. Mahnert attempted this through Switzerland. Radio operator *Ogfr* Volkmar reported via radio message, "right engine unusable – countermarch as ordered through Switzerland – Emergency bomb jettison," which meant that a flight through Switzerland was included.

The badly hit G1+HS flew over the Swiss border at circa 1,700 m altitude at 0939 hours by Bernex.

Factory No. 1705 originated from the Norddeutsche Dornierwerke.

The German pilot ignored the landing demands of the Swiss air raid defenses. The consequences were fatal.

During the overflight of the Geneva airfield, the Germans were requested to land by means of a green signal rocket from the *Platzkommando*. When no reaction followed, a platoon of the light Flab Battr 38 opened fire, without impact, however.

A *Patrouille* of *Fl Kp* 15 with a resistance assignment over the Pruntrut was set on the bomber. Over Yverdon they saw the He 111 at a low altitude, and moving into an attack. After a nose dive from 5,000 m *Hptm* Lindecker fired, followed by Lt Aschwande from circa 100 m with the cannons on the He 111. Resistance fire was not noticed. *Gefr* Lindner, the aerial gunner, was severely injured during this attack. The right aileron of the He 111 was so damaged that it was only controlled with difficulty. Mahnert turned towards the south, and was able to make an emergency landing by Ursins in a grain field.

A platoon of *Füs Kp* II/15 was immediately on the scene and disarmed the crew. Flight mechanic *Uffz* Schubert attempted to set the aircraft on fire by means of a Thermit incendiary agent. However, he was prevented by the forceful interference from a *Leutnant*, who received a minor injury from a shot to the head.

While the pilot and the lookout were handed over to the intelligence service for questioning, the three injured crewmembers were brought into Yverdon hospital. The aerial gunner died a short time later due to injuries suffered during the attack of the Me 109.

The He 111 was approximately 30 percent damaged. In the following days it was disassembled and transferred to the Buochs airfield via street, rail, and see route. For the first time in Switzerland one had the opportunity to inspect a modern bomber, as well as its equipment. The assessment of the damage revealed that circa 10 hits from 20 mm cannons and 12 more from MGs arose.

The German crew put on record that Spitfires above France attacked them. These statements were also taken by the Swiss bureau of investigation, and found again in all reports.

Gc II/7, with the Dewoitine D.520 stationed in Marey-sur-Tille, reported an attack on a He 111 on the morning of 2 June that subsequently crashed in Switzerland. Confusion between the Spitfire and D.520, which are quite similar, had to be thus assumed. Because the Swiss Me 109 only shot with cannons, the mentioned twelve MG hits must be attributed to the D.520.

The order for the Swiss *Jagdpatrouille* to attack aircraft of warring states without forewarning was certainly problematic, particularly when it dealt with an aircraft with combat damage and injured crewmembers.

The shooting down of the He 111 incited great outrage at KG 55. *Uffz* Mahnert, however, did not want to follow the landing order above the Geneva airfield. Within his crew this led to a discussion. While one was for a landing in Switzerland, Mahnert asserted himself for a continued flight to Germany. This was also the reason why the B-Stand was not occupied, and the crew was completely surprised by the attack of both Messerschmitts.

An Me 109 E of Fl Kp 15 above the Jura range.

Hptm Lindecker and Lt Aschwanden could identify from a circa 100 m distance ten hits with FF-K. The aerial gunner of the He 111 was thereby severely injured.

A photograph with an international character. The German He 111 on a flat wagon of the Belgian State Railroad and a Swiss ship on Lake Lucerne.

The transport of G1+HS through the narrow streets of Buochs. The aircraft was again given to the German authorities after the inspection.

4 June, Göring's Revenge
It was anticipated that a reaction would follow from Germany; within a few days the *Luftwaffe* lost five He 111 over Switzerland.
The aircraft was one thing. The numerous dead, injured, and interned crewmembers was, however, a high price for several border violations with the small neighboring country.

On the evening of 3 June the *Kommandant* of the V. *Flieger-Division*, *Generalmajor* Ritter von Greim, collected in Lachen-Speyerdorf the officers of II. *Gruppe* of ZG 1 equipped with Me 110s.

He spoke of a special assignment that would be aimed at flying along the Swiss border to determine if the Swiss would also attack the *Zerstörerverband*.

Oblt Brossler from KG 55, who did not simply want to accept the shooting down of the stricken He 111 of *Uffz* Mahnert, offered himself as a decoy. The undertaking was declared as a "reconnaissance flight with destroyer protection." However, harming the Swiss air sovereignty was to be avoided.

On Tuesday, June 4, 1940, due to the daily border violations, various *Alarmpatrouillen* were again placed on high alert. During the early morning, numerous violations of Swiss airspace were already occurring.

The destroyers of II./ZG 1 attempted an initial provocation between 0930 and 1100 hours, in which they flew criss-cross over the Pruntrut without being bothered.

In the afternoon, at approximately 1445 hours the border violations reached a new high point: the Swiss *Jagdpatrouillen* until now had dealt with single foreign aircraft, but they were now involved in real air combat with a superior force.

A C 35 (C-137) of Cp av 2 was located over the Dent de Vaulion, south of Vallorbe, on a *Patrouille* flight as the crew became aware of air combat between French and German *Jäger* in the Pontarlier and Frasne area. The German *Jäger* flew in a circle, and were gradually forced in the direction of Switzerland. The bold pilot wanted to resist the Germans with his venerable biplane, but was besieged from his side by three Me 110s and turned away. When a single He 111 became visible he went to attack, and fired from circa 700 m with

J-329 of Oblt Rufer after the off-field landing in Biel-Bözingen on 4 June. Notice the bullet holes on the propeller, the front section, and the water cooler.

the 20 mm FM-K and the MG 29 on the bomber without being able to determine impact. The German aircraft subsequently disappeared into the clouds. Lt Morier, without knowing, came upon the unit that Göring had sent out to Switzerland for punitive action.

The entire II./ZG 1 and one (or several) He 111 of KG 55 flew back and forth on the border near Neuchâtel-Jura.

The Doubs, clearly recognizable in this region as the line of the border, showed that the Germans again and again invaded Swiss territory with the intent of provoking the Helvetic *Jäger*. They did not hesitate long, and a series of heavy air combats followed. Thereby the Swiss pilots were for the first time confronted with, for them, an unfamiliar combat technique. As soon as attacked, the opponent formed a socalled resistance circle. This meant when they positioned themselves behind the opponent within the circle, they came inevitably into the field of fire of a following aircraft.

The records of the single actions are contradictory. On one hand it is dates, while on the other, the identification of the aircraft pattern contradicts.

On 4 June 4 Fl Kp 9 was in a combat mission with a Patrouille from Lausanne for the first time.

J-308 is prepared for action. Fl Kp 15 was equipped with Me 109 D and E. The mission of the "Jumo" led to a catastrophe on 4 June.

II./ZG 1 was equipped with Messerschmitt Me 110s. In the combat reports of the participating pilots, however, only "Bomber," "Heinkel-Bomber," "German Heinkel," "He 111," and "Do 17" were addressed. Only *Oblt* Borer of Fl Kp 14 recognized the Me 110, which was obviously unfamiliar to the Swiss pilots. It is not identified in the available documents how many He 111s actually participated in this punitive action.

According to the first reports at approximately 1422 hours, regarding foreign aircraft over the Neuchâtel-Jura, an entire series of scrambles of the Messerschmitt and Morane units took place.

Oblt Kuhn and Lt Aschwanden from *Fl Kp* 15 found themselves at 1500 hours in the St-Ursanne region when a report on "four bombers" in the Neuchâtel region was received. Above St-Blaise they saw the unit, but lost one another. After an attack from Lt Aschwanden, the supposed bombers turned towards the West and left Switzerland.

Lt Aschwanden was attacked by two Me 110s and had to abort the attack. After turning left he pulled high, and could then pursue a Me 110 until St-Ursanne. Aschwanden attacked for the second time, and opened fire from circa 300 m with the FF-K. The shots were in the direction of the aircraft, however, the German was able to escape over the border.

The German "bombers" flew along the Doubs at the same time, during which they constantly violated Swiss airspace. With this action they attempted to decoy the Swiss into French airspace.

An aircraft abruptly detached from the unit, and provokingly flew in the Pruntrut. *Oblt* Kuhn attacked the German over St-Ursanne. The tail gunner opened fire from a distance of 400-500 m. Kuhn kept his nerve and fired from a distance of 250 m at the aggressor so long that the FF-K was out and the MG 29 blocked. The aircraft, possibly a He 111, escaped—hit multiple times—in a steep flight in the direction of France.

The Me 109s from Kuhn and Aschwanden were damaged during the attacks by resistance fire from the aerial gunners, but they returned without problems to their base in Olten.

Lt Aschwanden immediately made a report on the situation to the acting *Kp Kdt Oblt* Rufer.

Shortly thereafter (it was 1519 hours), the *Patrouille* "Ursula 7" took off with *Oblt* Homberger and Lt Egli. *Oblt* Rufer followed at 1520 hours, and closed ranks with the *Patrouille* "Ursula 7" in the Solothurn region.

Oblt Homberger on a "Jumo" had to return back to Olten due to technical malfunctions. Rufer and Egli continued the flight in the direction of La Chaux-de-Fonds. *Oblt* Rufer sighted eight Me 110s over L Chaux-de-Fonds that formed a circle, and positioned themselves behind the last aircraft.

Shot on the left fore flap of Oblt Rufer's J-329. Notice the uncovered opening for the MG 17 (installed in the Me 109 E-1).

The German tail gunner opened fire from approximately 300 m distance, and caused several impacts. *Oblt* Rufer shot aflame the right engine of the Me 110 from approximately 200 m distance, but had to turn away because he discovered opposing aircraft behind him. Because the weapons were empty he decided to fly back to Olten. The water cooler of his aircraft, however, was damaged by impacts of bullets. Because the water temperature suddenly displayed 120 degrees, Rufer decided to land in Biel. Upon examination, further impacts in the engine and propeller, as well as on the wings, were determined.

Rufer's victim, a Me 110 of 6./ZG 1, crashed at 1550 hours in French territory in Grande Combe des Bois, by Le Russey. The pilot, *Uffz* Killermann, and the radio operator, *Uffz*. Wöhl, thereby lost their lives.

Lt Egli likewise fired on the Me 110 of *Uffz* Killermann, but could not determine a clear result, and had to abort combat due to weapons failure.

Lt Aschwanden, for the second time in action, attacked a Me 110 over the Chaumont that set off in the direction of Bielersee. During the pursuit of the German he was attacked from behind, and had to abort after several impacts.

After that the Germans flew in the direction of the border, and Lt Aschwanden returned back to Olten.

The *Morane-Patrouille* of *Flieger Kp* 13, *Oblt* Mark Wittwer (J-24), and Lt Ewald Heiniger (J-34) took off at 1503 hours in Biel-Bözingen for resistance flight in the Saignelégier region. Twenty minutes later they interfered in the combat over La Chaux-de-Fonds. It was the first combat mission ever with a D 3800.

The observations of both Morane pilots were interesting. During action they saw an aircraft turning around the longitudinal axis steeply crash. This was probably the Me 110 of *Uffz*. Killermann.

Another aircraft, with a smoke trail *steeply piercing* towards the North over the border, was presumably the He 111, which was attacked by *Oblt* Kuhn.

The *Patrouille* of *Fl Kp* 14, with *Oblt* Borer and Lt Ris, were at the same time over the *Freibergen*, and had no contact with German aircraft.

For the first time, *Fl Kp* 9 was also in a combat mission. A *Patrouille* from Lausanne interfered in combat in the La Chaux-de-Fonds region. *Hptm* Hitz (J-332) could maneuver into an attack position, and fired at an opponent with the cannons without determining impact. Lt Nipkow (J-351) did not make any shots. Both had to abort the attacks due to lack of fuel, but were able to land securely in Avenches.

The mission of the *Patrouille* Oblt Suter (J-308?) and Lt Rickenbacher (J-310) ended tragically.

All DB aircraft of *Fl Kp* 15 ready for flight were already in action. The *Patrouille* took off at approximately 1535 hours from Olten in Jumos. Both aircraft did not have radio equipment installed.

Rickenbacher, whose aircraft was the first ready for flight, took off in the direction of Saignelégier. Suter caught up after circa 5 minutes, and both flew further with approximately 100 m interval in *Patrouille*.

Lt Rudolf Rickenbacher, fallen on June 4, 1940.

The location of impact of J-310 by the Boécourt Cemetery. Lt Rickenbauer was located circa 400 m next to his aircraft. A simple memorial stone today reminds of the event.

Suter found himself at circa 3,500 m altitude over Saignelégier when he recognized three aircraft in the La Chaux-de-Fonds region that flew on the other side in a northeast direction. Because the line of the border was not clearly recognized due to the clouds, he remained at a distance. At this time *Oblt* Suter had lost sight of his *Patrouillen* comrades.

When the three Me 110, identified as "bombers" in the combat reports, became visible again, a single aircraft was before it. It turned suddenly on its back, and flew into the blanket of clouds in a steep gliding flight. The exact location could not be determined because of the clouds. Presumably, the single aircraft was the J-310 of Lt Rickenbacher. The fact is, the J-310 crashed nearly vertically with a running engine, and rammed into the ground near Friedhof von Boécourt. Rickenbacher crashed after a free fall with a torn parachute circa 400 meters from his aircraft. The exact circumstances of the deadly fall of Rickenbacher are not perfectly explained.

The investigation of the incident revealed that the J-310 caught on fire, and Rickenbacher was hurled from the aircraft in a brusque maneuver. It remains unclear why the canopy was released, the seatbelt torn, and the parachute prematurely opened.

The death of Rudolf Rickenbacher incited great outrage through the country. On 7 June, at the burial in Lotzwil, it brought forth an embarrassing incident: to honor the fallen Swiss, Göring had a wreath laid down, which was seen as a provocation from the mourners present. The outraged crowd tore up the wreath after the event.

The Me 109 D "Jumo" was no longer utilized for combat missions after 4 June.

The German *Luftwaffe* had to pay for this action again with two dead and many injured flyers. A Me 110 of 6. *Staffel* went missing, while others were more or less heavily damaged. An He 111 from KG 55 acting as a decoy had to make an emergency landing south of the Black Forest with an injured crewmember.

Over the course of the night of 6 June, another unidentified aircraft of the German *Luftwaffe* was shot by anti-aircraft, and crashed on the other side of the border.

These actions, for the fist time, had a diplomatic aftermath. The Reich's government intervened in Bern with a note, in which it protested the "hostile acts," and the "unparalleled operations of a neutral state." It further demanded that the *Bundesrat* apologize for the "unheard of incidents," and make up for the resulting damages.

On June 8, 1940, a C 35 of Fl Kp 10 was shot down during a border surveillance flight. The occupying troops were thereby killed. Already on 1 June a C 35 of Cp av 2 was involved in combat with aircraft of the German *Luftwaffe*.

Members of Fl Kp 15 in front of Homberger's J-328. Right next to the Swiss Cross is Wm Hans Ruegge. As of 1941 he was an instructor with the *Fliegertruppe* in Payerne.

A threat from Berlin that one would proceed with military action against Switzerland if necessary was taken seriously by the *Bundesrat*. Nevertheless, one had to repudiate the German reproaches, and insist on the right of a neutral state to defend its air sovereignty by all means.

When the response note from the *Bundesrat* was given to the German envoy on 8 June, the German threats were already put into action.

June 8 - The Boundary is Overstepped
On Saturday morning, 8 June, a C 35 (C-125) of Fl Kp 10 was on a border surveillance flight in the Pruntrut region.
At the same time six Me 110 flew over the Ajoie. Two aircraft abruptly detached from the unit, and shot down the C 35 by Alle. The pilot, Lt Meuli, and the lookout, *Oblt* Gürtler, did not have a chance. The attack came so surprisingly that they could not even activate the weapons.

The news of the death of both flyers spread like a wildfire. The *Kompaniekommandanten* did not at all await the mission order, and on their own initiative took off with all aircraft ready for flight. Between 1150 and 1210 hours seven *Messerschmitt-Patrouillen* rose to the resistance of the German aggressors:

Fl Kp 15 from Olten
- *Hptm* Lindecker / Lt Egli
- *Oblt* Homberger / *Oblt* Kuhn

Fl Kp 21 from Dübendorf
- *Hptm* Hörning / *Oblt* Borner
- *Oblt* Streiff / Lt Mühlemann
- *Oblt* Scheitlin / *Oblt* Kislin and *Oblt* Köpfli
- *Oblt* Willi aborted the mission due to engine failure, and returned back to Dübendorf

Cp av 6 from Thun
- *Oblt* Hadorn / Lt Thurnheer
- *Oblt* Liardon / Lt Benoit

A total of 15 Swiss Me 109 Es, and presumably 28 German Me 110 Cs were opposing each other.

On the eve of 8 June *Generalmajor* Ritter von Greim stood before the pilots of II./ZG 1 in order to proclaim a punitive action against Switzerland. This time a bomber as a decoy would be done without. Additionally, II. *Gruppe* of ZG 1 was ordered for action.

The Germans used the same tactic as on 4 June, in which they attempted to decoy the Swiss in a defense circle. 5./ZG 1 under *Oblt* Schmidt flew at 2,000 to 4,000 m altitude in triple formation, like bomber units. The *Stabsschwarm* under *Hptm* Dicoré, and 4./ZG 1 under *Hptm* Kaldrach, circled this supposed bomber unit at the same altitude, and 6./ZG 1 under *Oblt* Kadow had the assignment to cover the unit at a 6,000 m altitude.

Oblt Homberger had to make an emergency landing on 8 June with his damaged aircraft in Biel-Bözingen. 34 bullet holes were counted.

The absence of head and back armor led to severe injuries of the pilot. The KTA was of the opinion until the end of 1944 that no armor plates could be installed in the cockpit of a Me 109.

Oblt Homberger in the hospital in Biel. Rudolf Homberger was a popular top athlete, and participated in the Olympics in Berlin in 1936 as a rower in the eight.

Picture above: Oblt Kuhn forced back the pursuers of Oblt Homberger and shot down a Me 110. On February 21, 1970, Hans Kuhn lost his life as a passenger of SWISSAIR "Coronado" HB-ICD during an attack by Palestinian terrorists in Würenlingen.

J-328 was disassembled in Biel-Bözingen by the *Armeeflugpark* and brought to Buochs for repairs. The aircraft had 74.55 operating hours.

The battles that played out are difficult to ascertain. The attacks were flown individually, or in the *Patrouille*. It cannot be determined with certainty who combated whom, or shot down or damaged an opponent.

Fl Kp 15, from the nearby Olten, was the first to attack. The *Patrouille* Lindecker/Egli flew from the direction of the sun in the uppermost defense circle of 6. *Staffel*, and was able to shoot down the Me 110 of Jochen Schröder, who had to abort combat with a burning right engine.

The *Kommandant* of 6./ZG 1, *Hptm* Kadow, was wounded during an attack, and his aerial gunner, *Uffz* Wunnike, was killed by a shot in the head. In the list of losses of II./ZG 1, the aircraft was listed as a total loss. It is unclear who Kadow attacked, and where he came to ground. Morteau (F) was named as a possible landing location.

The *Patrouille* Homberger/Kuhn attacked 4. *Staffel*, which formed a resistance circle around the "bomber unit." *Oblt* Homberger was caught in the attack fire of a Me 110, and was thereby seriously injured.

He attempted to reach Olten in low level flight, but was suffering from impaired vision due to his injuries. Homberger decided to land in a direct approach in Biel-Bözingen. During landing the left landing gear buckled, because it did not lock due to a shot oil tube. After a Cheval de bois the aircraft came to a stop.

The severely injured pilot climbed out of the cockpit by himself, and afterwards collapsed unconscious.

Members of *Fl Kp* 13 brought Homberger to Biel hospital; he had bullet injuries in the back, lungs, and pelvis. 34 bullet holes were counted on his aircraft.

Oblt Kuhn saw how far below Homberger was besieged by many Me 110. In a nosedive, he reached a favorable shooting position and fired with both MG 29 on the rear aircraft, but it immediately turned away over the wings.

Due to increased speed Kuhn was immediately behind the second Me 110, which was positioned for an attack on Homberger. Homberger's aircraft constantly lost altitude, and trailed a whitish-gray cloud of smoke behind it. Kuhn came to a close distance, and fired on the opponent with FFK and MG 29. The aerial gunner of the Me 110 obviously had problems, because there was no resistance fire. At the height of the sport airfield, Courtelary Kuhn noticed that behind him four Me 110s were moving into an attack position. In neck-breaking zero altitude flight through the Taubenlochschlucht, he succeeded in getting rid of his pursuers—he could assume that no foreign pilot would dare to follow him into this area.

The crew of the Me 110 did not have any luck. After the last attack by *Oblt* Kuhn they barely came over the mountain hollow by Pierre Pertuis, flew over Tavannes towards St-Ursanne, and crossed over the border between Réchésy and Pfetterhouse.

J-346 of Fl Kp 15 with a damaged landing flap.

Oblt Streiff, Fl Kp 21, shot down a Me 110 of 4./ZG 1 at Triengen. Both crew members were killed.

In the region of Sepois-les-bas, Fw Breiter could no longer maintain his aircraft, and crashed west of Überstrass in a fish pond in the middle of a forest. Radio operator *Ogfr* Hink was able to escape the aircraft shortly beforehand with a parachute and was taken captive. The pilot lost his life in the crash.

The *Patrouille* Streiff, Köpfli, and Scheitlin of *Flieger Kp* 21 was likewise involved in a skirmish with 4./ZG 1 *Staffel*. Meanwhile, the combat stretched from Chasseral to Laufental.

In the area of Zofingen, the *21er* forced away a Me 110 from their unit. By Triengen, *Oblt* Streiff succeeded in cutting off the path of the opponent in a turn, and bringing him to a crash through the firing of both MG 29. Aerial gunner *Obergefreiter* Hoffmann attempted to jump with a parachute. However, his parachute was released too soon, and caught on the horizontal stabilizer. The pilot, *Unteroffizier* Scholz, also lost his life in the crash. The aircraft rammed into the ground by Weiler Wellnau. The onboard clock displayed exactly 1255 hours.

Oblt Borner, *Fl Kp* 21 (J-364 without radio equipment), lost the *Staffelkommandant* Walo Hörning over the Pruntrut in a cloud. Subsequently, he climbed to 7,000 m altitude in order to orient himself.

Later he said, "It looked like the inside of an aquarium—aircraft at each altitude, all circling around to the left... Kahn on his head and into this tower."

After several attacks, Borner's Me 109 was also hit by a tail gunner. This resulted in the blockage of the left aileron. Steering with the rudder, Borner flew southward over the Jura range with the intent of landing on the airfield in Biel-Bözingen. But at the same time, the aircraft of *Oblt* Homberger blocked the runway there.

Again in the accompaniment of *Hptm* Hörning, Borner flew to Olten, the base of *Fl Kp* 15, where he landed safely.

When the *Patrouillen* of Cp av 6 from Thun arrived over Saignelégier, air combat was already in full swing. According to statements of the pilots, they attacked aircraft that flew in triple formation several times.

The numerous weapon failures in Cp av 6 were remarkable. *Oblt* Hadorn could only fire a few shots with the cannons, as they were blocked. Because the left MG 29 also had a loading failure, only one weapon was left for the combat mission. Both cannons jammed one after the other on the aircraft of Lt Benoit who, however, could continue to fly his attacks with the MG 29. Lt Thurnheer flew his mission with a jammed left MG 29. Only *Oblt* Liardon had no problems with the weapons and could make more hits on a Me 110 over Delémont with cannons and MG.

In the last phase of combat, only single aircraft were involved. The German formation was long since driven apart, and single aircraft of the 5. *Staffel* found themselves over the Laufental.

The Me 110 C-1 (2N+GN) of 5./ZG 1 that made an emergency landing at Nunnigen.

111

Pilot Fw Dähne and aerial gunner Ogfr Klinke (with a cast). Both tried to satisfy the Bureau of Investigation that they were on an escort mission in the Belfort area. However, the entire II./ZG I called for punitive action against Switzerland.

The Me 110 (2N+GN) was attacked in the Laufen region by Me 109 of *Fl Kp* 15, which managed several hits. The aerial gunner returned fire. Shortly thereafter, by Breitenbach, the aircraft got caught in the fire of a 7.5 cm battery of *Flab Det* 80 and was further damaged. Afterwards they were shot from Büsserach by four MGs of *Mitrailleur* KP IV/26. The distance, however, was too great to make an impact.

The pilot set his damaged aircraft on its belly by Oberkirch/Nunningen, whereby the aerial gunner broke two fingers.

At first, both unsuccessfully attempted to disassemble the MG 15; they wanted to set the aircraft on fire with a lighter. Subsequently, both fled in a southern direction. However, after a short amount of time they were caught by soldiers and countrymen and led back to the aircraft where, in the meantime, the police also arrived.

The pilot, *Fw* Dähne, and the aerial gunner, *Ogfr* Klinke, had to later unload the weapons from the aircraft and deposit the munitions outside of the aircraft. The Me 110 C-1 Factory No. 2831 was only slightly damaged, and attracted great interest from the military experts.

On the evening of this eventful day, on both sides of the border one went over the books. Switzerland lamented two dead and a severely injured pilot. The German *Luftwaffe* again had to accept four dead, along with several injured and interned crewmembers. Regarding aircraft, a C 35 and four Me 110 went missing, while further aircraft were more or less badly damaged.

Diplomacy came into play, and between Berlin and Bern the wires were running hot.

While in the *Fliegerkompanien* victory celebrations were being held, and the Swiss population was rejoicing their "heroes of the nation," the responsible politicians and military saw themselves increasingly subject to the pressures of the German Reich government.

Still on the evening of 8 June, *Kdt FF* Trip ordered to abandon each air battle 5 km before the border. Thus, for the Swiss *Jagdflieger* an entire series of limiting measures began. On 10 June the border surveillance flights were abolished. The General decreed a prohibition for air combat over the Ajoie on 13 June. Foreign aircraft could only be attacked in self defense, and insignificant border violations were no longer reported. Finally, on 20 June came the complete prohibition for air combat, and action of military aircraft over the entire Swiss sovereign territory. The neutrality protection devolved completely to terrestrial anti-aircraft.

The Swiss *Jagdflieger* uncovered the defects of the twin-engine Me 110. As heavy *Jäger* they had only few chances against the agile Me 109.

On the evening of 8 June, the combat operations for the Troop were over. The Swiss *Flugwaffe* had passed their trial by fire. Now it was up to the diplomats to not let the subject escalate. Olten, Fl Kp 15.

The history showed that these measures were correct, because no one could predict how the powers in Berlin would react to an escalation of events. Germany was in combat with hundreds of aircraft against England and France. The aerial battle over Dunkirk had just ended, and the German *Luftwaffe* could fully concentrate on the further actions in France.

A massive annihilating blow against the Swiss *Flugwaffe* could, with certainty, have not been avoided.

Furthermore, with Italy's entry into the war on 10 June, Switzerland became an island in the middle of Europe.

However, the members of the *Fliegertruppe* felt sold, when on 1 July Bern officially apologized to Germany for the air battles. On 26 July personal instructions of *Kdt FF* Trip, *Div* Bandi, prohibited "unobjective" discussion on the "flyer incidents," and the mood sank to a low point. Switzerland set still further signs of easing tension.

According to the Haager Convention from 1907, the interned crewmembers were permitted only after the end of hostilities to be led back to their home. England, as an ally of France, continued to find itself in a state of war with the Third Reich.

Already at the end of June 1940, the German air crews could leave Switzerland. Also, the *Luftwaffe* members fallen over Switzerland were transferred to Germany. The aircraft that were shot down or made an emergency landing over the Swiss sovereign territory during air combat, or respectively their debris, were given back to Germany until the year's end.

The balance sheet in the operational area at Flieger Kp 15. Left, the portrait of Lt Rickenbacher. Center, various writings of congratulations from the population, and the type plate of Ursin's He 111. Right, a photograph of Oblt Homberger.

Events in May-June 1940

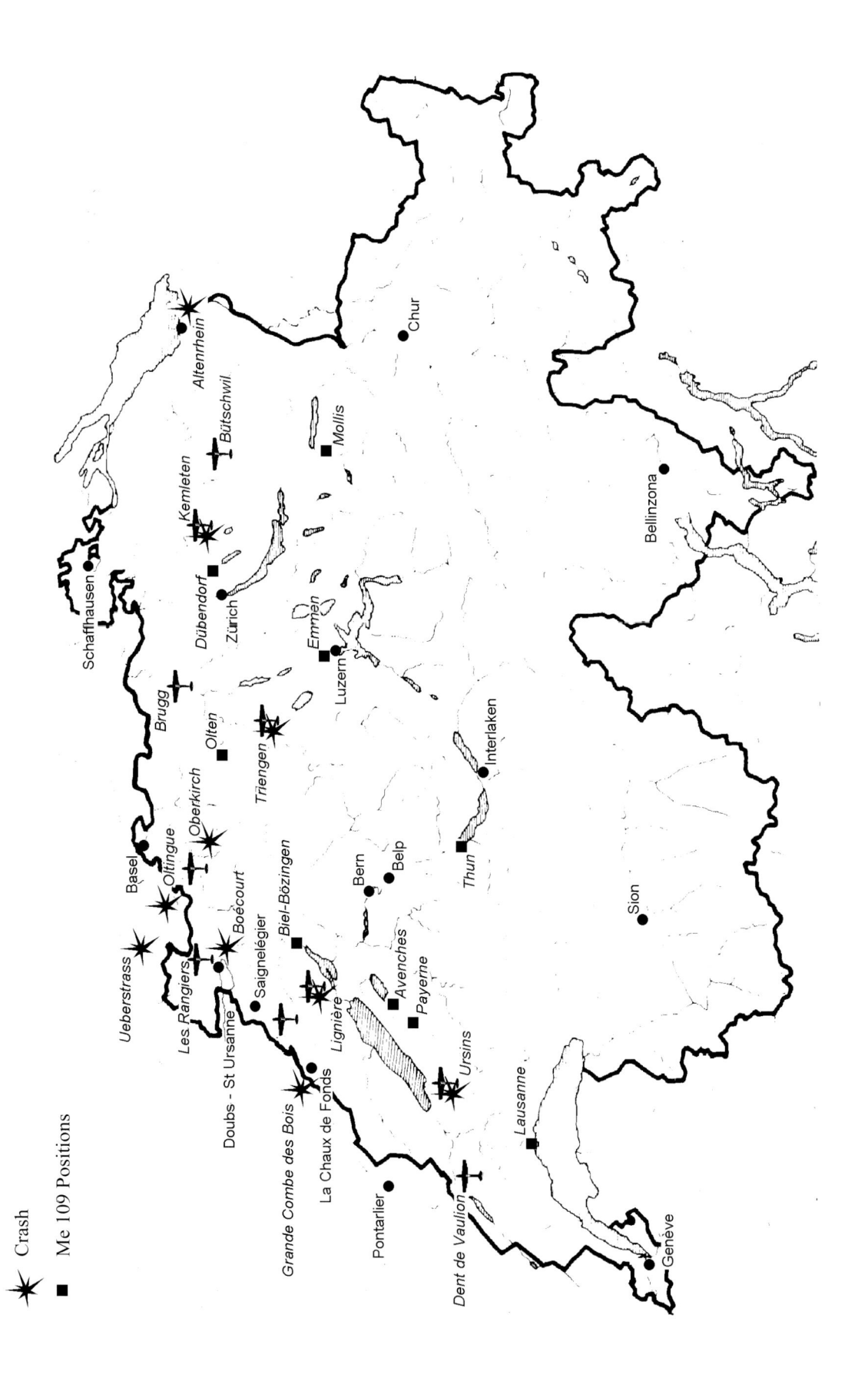

Air Combat — ✈
Crash — ✵
Me 109 Positions — ■

114

Chapter 12:

The Events as of Summer 1940

Flyers with Propped Wings

After the events in Summer 1940, it became calmer in the Swiss airspace. The action was, for the time being, primarily occurring over England.

The following months in the troop on duty were often characterized by monotony. During this time, in order to increase motivation, the painting of *Kompanie* emblems was allowed. These were more cheerful than the aggressive emblems normally read on the aircraft during the length of duty, and then were to be again removed.

At this time in *Fl Kp* 15 a spectacular espionage case occurred. The *Einsatz-* and *Nachrichtenoffizer Oblt* Reimann used his rank to provide the German intelligence service with all kinds of information. Thereby, members of other branches of military service supported him. The matter leaked out, and during his arrest no fewer than 176 topographical cards with notes were found. Reimann and two of his assistants were sentenced to death, and executed for treason.

The expansion of the war to the Balkans and eastern Europe in summer 1941 had no direct consequences for Swiss air sovereignty. However, the RAF increasingly carried out night attacks on targets in Germany and Italy, whereby Switzerland was also overflown.

The anti-aircraft that were authorized for the defense of the airspace could not prevent this. In September 1943 the anti-aircraft had less than 100 searchlights, and only 7 sound detecting devices available. Electronic location devices were not at all available.

The *Jagdfliegereinheiten* that were in rotation in active duty were on normal standby, which meant that, should the situation change, within two hours they had to be ready for action.

Scrambles and interception maneuvers were carried out only as training for the maintenance of operational readiness. Idleness for days, and over months, beat down the morale of all men in service.

The fact that foreign aircraft could use Swiss airspace practically undisturbed, and the *Flugwaffe* was not permitted to intervene,

After the events of June 1940, the neutrality protection was transferred to the air defense. 7.5 cm Flab Kan 38, Schneider type from France.

was seen from the population as well as the troops as a weakness of the military and civil leadership.

On January 1, 1943, a 4. *Fliegerregiment* was deployed. The *Messerschmitt-Einheiten* were subsequently reorganized:

Cp av 6 under Capt Liardon was in the purely French-speaking *Régiment d'aviation* 1.
Flieger Kp 7 *Hptm* Läderach, *Fl Kp* 8 *Hptm* Fischer, *Fl Kp* 9 *Hptm* Rufer, and *Fl Kp* 15 under *Hptm* Lindecker belonged to *Fl Rgt* 2.
Fl Kp 21, under *Hptm* Streiff, remained in *Fl Rgt* 3.
Flieger-Regiment 4 and the newly deployed *Überwachungsgeschwader* did not include any *Messerschmitt-Einheiten*.

The airspace violations during the day were primarily unarmored aircraft of the Axis powers. However, this changed in Summer 1943 when the USAAF began to fly their missions from England and North Africa.

The first attack on Austria (then called Ostmark) brought the first U.S. aircraft on Swiss ground. On August 13, 1943, the 9th USAAF and its units of the 8th USAAF from Bengasi, in Libya, took off for an attack on the Viennese Neustadt aircraft factories.

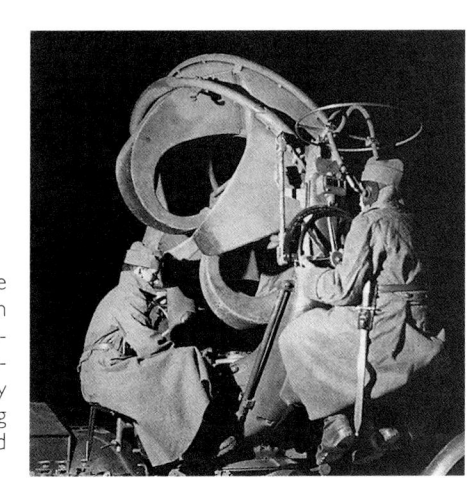

There were no available radio communication measuring devices (RADAR) available in Switzerland. One managed by means of sound detecting devices. "Elaskop" sound detecting device.

For an efficient defense at night the necessary number of of searchlights. "Galileo" spotlight

115

Flyers with clipped wings. After the ban on protecting their airspace with aircraft, the service was of a monotonous nature. Only in October 1943 were the Jadgpatrouillen permitted to be brought into action. Fl Kp 7, Avenches, July 1940

In the WNF at this time, the manufacture of the Me 109 G was running full blast. The attack came as a complete surprise, and resulted in a loss of a third of the monthly production due to the damages on factories 1 and 2.

The B-24 D "Death Dealer" of the 93rd F reached the target area with only three engines, where another was damaged by flak fire. Because a return flight to North Africa was not possible, 1st Lt Geron decided to reach Switzerland over—or more precisely through—the Alps. The daring flight succeeded: at circa 16:5 hours the Liberator landed by Wil/SG on an open field. The crew set the aircraft on fire, and were later interned.

The Reintroduction of the *Jagdpatrouille*
When Allied aircraft in increasing masses violated Swiss airspace also during the day, the prohibition to deploy *Abfangjäger* for protection of neutrality was raised in October 1943. By day three *Alarmpatrouillen* were maintained on three various airfields ready for action.

During the night, the *Flugwaffe* could not be deployed as before due to unfit equipment and means of management.

On a trial basis, night missions were flown by pilots of the UeG. Because no aircraft had a radio measurement device at their disposal, the mission was only possible in collaboration with flak searchlights in a restricted sector. Thus, the chances of success were correspondingly minor. This meant that the bombers of the RAF could continue to fly practically unhindered over Switzerland.

The Allied landing in the Gulf of Salerno in September 1943 brought a new situation for Switzerland on the Southern border. After the deployment of tactical combat aircraft in the Foggia region, the attacks against Southern Germany and Northern Italy increased. Concerning the border violations, if it involved single Allied aircraft that lost their way, or at the most wanted to escape enemy resistance, the Swiss air sovereignty was deliberately disregarded. But the Swiss air defense was powerless against the high flying bomber formations and fast reconnaissance aircraft.

The deployment of the *Alarmpatrouillen* was subsequently limited to the interception and escort of single, dispersed, or needy aircraft. The instructions to attack foreign military aircraft without prior warning were raised because an increasing number of aircraft with combat damage and injured crewmembers were flown into Switzerland.

In Summer 1940, to raise morale the display of Kp emblems was permitted. Fl Kp 7, Avenches, Summer 1940.

116

J-366 on a patrol flight in 1941. Oblt Dannecker, Fl Kp 8.

The continual airspace violations demanded the reintroduction of the procedure to request an aircraft to land.

Because no international norms existed in what manner foreign aircraft were to be requested to land, all accredited *Militärattachés* in Switzerland were informed of the procedures of the Swiss *Jagdpatrouillen* in such a situation on February 2, 1944. As a general rule a *Doppelpatrouille* was deployed.

Single foreign aircraft were not directly attacked, but first warned as follows:

One or two aircraft approached the foreign aircraft from the side under constant wobbling of the longitudinal axis. The approach took place at an incline, so that the aircraft's longitudinal axis never aimed at the foreign aircraft.

The landing order was issued by shooting a green rocket. At speeds under 300 km/h the landing gear could additionally be lowered.

The landing order was once again repeated, and within a time that permitted the foreign crew to react correspondingly. After a second warning the attack was ordered.

If the foreign aircraft acknowledged the warning, the head of the *Patrouille* ordered one to two aircraft to lead the aircraft in front to the closest airfield. After reaching the airfield, the leading aircraft indicated the landing direction with their own landing.

Two or more aircraft of foreign nationality, according to the instructions, were attacked without prior warning.

It is another question if this orientation on the *Militärattachés* also reached the numerous stationed units all over Europe, and there again all crews.

Misunderstandings and overzealousness, however, could have fatal consequences. This caused a tragic event on April 24, 1944, when a Morane of the *Überwachungsgeschwader* shot down the B-17G "Little Chub" of the 384th BF over the Greifensee; Thereby, six crew members lost their lives.

After the Allied landing in Normandy in Switzerland, one feared the possible invasion, and on 10 June ordered a mobilization. Thereby the entire *Fliegertruppe* was deployed. On 18 July *Fl Rgt* 2, 3, and 4 could already be released.

The neutrality protection was incumbent on the three *Staffeln* of the UeG, and the Rgt av 1 with Cp av 6 as the single unit on Me 109 Es.

Through the quick advancement of the Allied troops, the border violations also drastically increased. The tactical formations of the USAAF, which attacked targets in Southern Germany, increasingly used the relatively secure airspace over Switzerland during the inward and outward flight. The aggressive *Jagdbomber* pilots respected no borders: on 8 September the Delsberg train station was fired at, and a train of the Solothurn-Moutier railway was attacked. Days later a freight train by Rafz, a passenger train by Weiach, and on 11 September the express train Zurich-Basel by Pratteln were attacked with bombs and armaments.

These attacks near the border were of such a short duration that the *Jagdpatrouillen* could not attack.

As of October 1943, the Jagdpatrouillen were again introduced for neutrality protection. Emmen, date unknown.

The Avenches Base in Summer 1940

Flieger-Abteilung 3 (Fl Kp 7, 8, and 9). The plains by Avenches were already in 1910 used by the Swiss flying pioneer Rene Grandjean and Ernest Failloubaz as an airfield.

General Guisan used the Me 108 Taifun for his inspection trips many times. Avenches, August 15, 1940.

Defilee in Avenches. *Flieger Abt* 3 was stationed in Avenches from April 1 until September 3, 1940. In the air combat above the Jura, Fl Abt 3 was involved only at the border. At this time they were still in formation.

The Swiss Stud Farm. (Harras Fédéral).
Left, the aircraft of Fl Kp 9, above, along the L'Arbogne, Fl Kp 7 and 8.

118

A Me 109 E of Fl Kp 9 ready for action. Notice the inserted crank handle for the starter.

Commando and pilot area of Fl Kp 7.

The radio station of Flieger Abt 3. Notice the pedal generator for the generation of electricity.

The repair tent of Fl Kp 8, and the barn that was used for larger inspection work.

Small inspection of a Me 109. This maintenance work had to be conducted after 40 operating hours.

An engine exchange on J 343 of Fl Kp 7

The parking of the aircraft in the old barn was problematic. However, the concrete floor also had advantages. In the background the Rep-Zelt of *Flieger* Kp 8.

Operational control on the landing gear during a period of bad weather.

120

In these days one concentrated mainly on a good camouflage. Fixed foxholes were introduced just in 1943. J-317 of the Patr.II, Fl Kp 8.

In order to bring the aircraft to optimal cover, wooden plans were built on the bank of the L'Arbogne.

J-335 of Fl Kp 7 is prepared in natural coverage.

J-360 of Patr. II is prepared for the mission. Several aircraft of Fl Kp 8 were decorated with flower motifs. In the photograph the "Narziss." The motif was applied to both sides.

Lt Benoit, Cp av 6, did his service with Fl Kp 7 in Avenches. During a landing that was too flat J-330 overturned. Despite the completely destroyed canopy, the pilot was rescued uninjured. Avenches, July 16, 1940.

Hptm Läderach was *Kommandant* of *Flieger* Kp 7 during active duty. On April 22, 1949, he crashed fatally in a D 3801 (J-191) on Lake Neuchâtel during a rocket firing flight. In the photograph is Walter Läderach as a pilot with Fl Kp 18.

J-360 "Narziss" of Fl Kp 8, Summer 1940.

J-358 of Patr. I, Fl Kp 8 in camouflage. In the background, the barn for repair and revision work.

Oblt Hans-Paul Häberlin, Fl Kp 9. The pilots of this unit wore white head caps at this time.

Self-made rearview mirror, Oblt Häberlin, Fl Kp 9.

Fl Kp 9 was stationed on the street by the main building of the Harras Fédéral.

The Everyday Life of the Messerschmitt Units

Provision operations for a mission. Fl Kp 15, Olten, Summer 1940.

Preparatory work on a VDM propeller, Fl Kp 15.

Ground crew of *Flieger* Kp 21.

Pilots of *Flieger* Kp 21, Summer 1940. Picture below from the left: with helmet unknown, Oblt Scheitlin, Oblt Haase (Tech Of), Oblt Meyner, Hptm Hörning, Oblt Ahl, Oblt Streiff, Oblt Wannenmacher, Oblt Borner, and with helmet unknown.

After a tire blow-out at landing, J-383 swung off and landed on its head. Thereby, the fuselage suffered 30 percent damage. The right wing was bent outward 30 degrees. Oblt Köpfli, Fl Kp 21, Buochs, February 14, 1941.

J-330 in Emmen during the warming up of the motor. The aircraft was newly constructed after a severe accident in July 1940 in Avenches.

Cp av 6 during replacement service in Spring 1941 in Thun. Kp Kdt was Hptm Roubaty, in June he fatally crashed with the prototype C 3603.

For many soldiers, active duty had material consequences. It especially hit independent recruits, farmers, and family men hard who rendered services lasting for months. The social contributions for the army were minimal at this time.

Christmas 1941 away at war. The celebration of Fl Kp 9 was accentuated by the *Kompanie's* own orchestra "Goldkometen."

The Pilots of *Flieger* Kp 9, Autumn 1942. From left: Kp Kdt Hptm Rufer, Lt Dumelin, Lt Gerber, Oblt Häberlin, Lt Brenzikofer, and Oblt Kilchenmann. Lt Urfer shot the photography in Wilderswil by Interlaken.

126

At air combat practice on April 15, 1942, two aircraft of Fl Kp 15 collided. Oblt Brügger landed J-359 in Thun with a damaged wing.

J-311 was flown by Oblt Reber, who likewise landed safely in Thun.

On August 10, 1942, two Me 109s again collided in the air. Flz J-365 and J-366 flew independently of each other in the Gerzensee area. As a result of poor visibility—due to a cloud—J-365 hit the propeller of J-366 with its left wing. Both aircraft crashed, but the pilots were able to save themselves with parachutes. J-365 at the edge of the Kronbergwald.

The pilots of Cp av 6. From left: unknown, unknown, Oblt Schenk, Oblt Hadorn, unknown, Oblt Zschokke, Hptm Liardon, Oblt Burlet, Oblt Thurnheer, and unknown.

As of the end of 1942, Cp av 6 rendered their replacement service predominantly in Zweisimmen.

As of summer 1940, the Me 109 D was still utilized only for training purposes. Illustration from the rules of procedures, "Bauvorschriften für Flugzeug- und Tarnzelte."

A splinter defense provided with camouflage nets. The improper use of such nets caused Kdt FF Trp in September 1944 to issue particular instructions on the "use of camouflage nets."

As of 1943 fortified foxholes (U43) were introduced. The foxholes were later provided with sliding doors. Buochs, circa 1943.

An aircraft foxhole U 43 camouflaged further with nets.

Girls, such as the "SLEEPY TIME GAL," took care of turbulent times in minds, but also in the airspace. Due to the countless overflights of the Allied bombers, the *Jagdpatrouillen* were in 1943 introduced again.

A B-24J (42-99813) of the 464th BG, 15th AF. The aircraft landed in Dübendorf on July 19, 1944, with engine damage.

The fighter aircraft of the Swiss *Flugwaffe* in Summer 1944.

During this period approximately 200 D 3800/01 were delivered. The Me 109 E / J-389 bore the Kp No. 21 on its fin. In the background the Gustav J-708 of Fl Kp 7

"Shoo Shoo Baby," a B-24H (41-29431) of the 44th BG, was intercepted by a Me 109 E on March 18, 1944, at the border, and was led to Dübendorf. The aircraft had several hits in the fuselage and tail assembly which, however, had no influence on flight safety. At landing 1,300 gallons of fuel were still in the tanks.

The 100th American aircraft that reached Switzerland was a B-17G (42-31074) on July 13, 1944, in Emmen. The name of the aircraft was "CAHEPIT," which meant "cannot help it." Living up to its name, it roughly flew past a F+W building, destroyed a catapult construction, and slid over the rail tracks. At the smashing of an overhead contact line mast the B-17 caught fire and burned. The crew walked away with minor injuries.

The Boeing B-17G "FRECKLES" (42-107092) of the 401st BG was intercepted by a Swiss Me 109 on July 31, 1944, in the Lake Constance area. The pilot, Lt Ossiander, despite engine problems, tried to escape, but had to reluctantly follow the *Jäger*. The crew was visibly delighted when they saw the Swiss aircraft and safely arrived in Dübendorf.

The famous pin-up girls of Alberto Vargas was found on numerous bombers and leather jackets of crews.

During the approach to Munich on October 4, 1944, the B-24H (42-52485) was hit hardly by anti-aircraft guns. The forward gunner, Sgt Lostiss, thereby suffered severe injuries. After further hits the "BROWN NOSE" slide-slipped approximately 3,000 meters before the pilot 2nd Lt Peskin brought his aircraft under control again. He succeeded in bringing the damaged B-24 over the Swiss border, where it was led by a Me 109 to Dübendorf. The badly injured crewmembers had to be brought to the Zurich canton hospital.

A mistake with consequences. Oblt Heiniger's J-324 after the attack by a USAAF Mustang of the 339th FG.

5 September and its Consequences

Since mid-August 1944 *Fl Kp* 7, equipped with the Me 109 G, was in active duty in Interlaken.

After an emergency landing due to engine failure on 2 September all Me 109 G were barred from air service.

On 4 September *Fl Kp* 7, with nine Me 109 Es, resumed air service, and were ordered to Dübendorf to support the UeG.

During the early morning of September 5, 1944, over 660 bombers of the 8th USAAF took off from England for Mission 605. Their targets were in Karlsruhe, Ludwigshafen, and Stuttgart. Over two hundred B-17Gs of the 3rd Bomb Division had the assignment to destroy the Daimler-Benz aircraft engine factories by Stuttgart.

Over the target area, the bomber "Blues in the Night" of the 390th BG was hit several times by the flak. The crew had to shut down engines 1 and 4, and a further engine likewise malfunctioned. Nevertheless, 1st Lt Gallager managed to fly the hit B-17 in the escort of two Mustang P-51 B/Cs of the 339th FG into Switzerland.

The unit crossed over the border in the Stein a/R - Eschenz region at 1107 hours.

South of Frauenfeld it was covered with 25 shot from 7.5 cm flab cannons of *Flab Det* 95 without being hit.

In order to evade the fire, the bomber turned off in the direction of Winterthur. Over Embrach, the lower defense position—the so-called "Ball Turret"—was released, which smashed through a house, but only did material harm. The time was circa 1120 hours.

At 1113 hours the Liberator "Lonesome Polecat" of the 389th BG flew over the border by Schaffhausen. The B024H-25-FO (42-95205) was a socalled "strike photo and radio contact" bomber. The aircraft belonged to the 2nd BD, and was damaged by the flak over Karlsruhe. With two defective engines it arrived at the Swiss border, escorted by two Mustangs, and subsequently flew further to Dübendorf without escorts. *Fl Kp* 7 in Dübendorf received the air raid warning at 1110 hours.

The pilots were set at alarm level 3/2, and were already in the cockpit. The initial instructions for 1. *Doppelpatrouille* stated. "Scramble, position Zurich, 3,000 m, surveillance towards the East."

At 11:12 hours the *Oberleutnants* Künzler/Schoch and Heiniger/Treu took off.

The *Kommandant Hptm* Läderach, who actually led the *Doppel-Patrouille*, had just returned from a test flight with the interned P-51B, and was surprised by the alarm.

The aircraft of his substitute, *Oblt* Heiniger, had radio equipment that did not function consistently, whereupon *Oblt* Künzler took over the leadership of the unit.

When the four aircraft were over Zurich, E.Z. Bern transmitted the following message at 1117 hours: "Aviso 1 bombo + 2 bibi positione Loki direzione Atlanta," which meant, "Attention 1 bomber + 2 *Begleitjäger* over Winterthur in Westerly direction."

The trigger of the affair of 5 September was the B-17G-70-Bo "Blues in the Night."

132

The site of the crash of Oblt Paul Treu by Zürich-Neuaffoltern.

The problem with broadcasting was once again displayed. *Oblt* Heiniger only understood the word "Loki," and *Oblt* Schoch the fragment "positione Loki," while the others heard nothing at all. *Oblt* Schoch caught up to the *Patrouillen* leader and turned in the direction of Winterthur, whereby the three followed him.

In the area between Winterthur and Kloten they saw the hit B-17 "Blues in the Night." Künzler and Schoch positioned themselves next to the aircraft according to the standard procedures in order to indicate the way to Dübendorf, while Heiniger and Treu followed at a distance of 500 m above and 500 m, respectively, 800 m behind the bomber. That the B-17 was under escort protection was furthermore unknown to the pilots. *Oblt* Künzler fired a green signal rocket, and thereupon flew in a left turn to indicate the direction of Dübendorf to the bomber.

The B-17 followed, but without acknowledging the signal. After Künzler indicated the landing direction to the B-17, he fired two more green rockets. The B-17 would have had to release the landing gear in order to signal the intent to land. What the Swiss did not know was that the 4-engined bomber flew with greatly reduced engine power, and the pilot did not want to cause additional resistance. When Schoch likewise fired a rocket, the crew of the B-17 reacted with the firing off of a *Doppelsternrakete*.

Neither *Oblt* Künzler nor *Oblt* Schoch noticed, however, the incident that was playing out at the same time behind them.

After the 1. *Patrouille* turned off in the direction of Dübendorf with the bomber, *Oblt* Heiniger observed the escort mission as complete, and turned left away with *Oblt* Treu. Shortly thereafter they spotted the B-24 "Lonesome Polecat" likewise flying in the direction of Dübendorf, however, not the two Mustangs that flew high above them.

1st Lt Erickson and 2nd Lt Ostrow of the 339th FG interpreted the turning maneuver of the two Me 109s presumably as an attack on one of the bombers, and immediately attacked the *Patrouille*.

Both Swiss pilots were completely surprised by the attack. *Oblt* Treu, fatally hit, crashed with his J-378 by Zurich-Neuaffoltern, into the Hürstenwald. Heiniger's Messerschmitt was hit during the first approach, but through ingenious maneuvering he was able to evade further attacks.

Oblt Heiniger set his smoking J-324 with its idle engine and retracted landing gear in Dübendorf on its belly, whereby he hit his head hard on the aiming device. With a minor head injury, bruises on his ribs, and burns on his right arm, he was brought into the *Kantonsspital* Zurich, but was released after three days.

P-51B 42-106438 contributed to the general confusion on 5 September. The aircraft of the USAAF 4th FG landed in Ems on 19 July, and was later utilized for comparative flights. Latest information states the number J-900 was never used.

133

The debris of J-378. The absence of an armor plate in the cockpit of the Me 109 also had fatal consequences for the pilots.

The attack of the U.S. fighters was observed from the Dübendorf airfield without any ability to somehow intervene. Many factors played out in these dramatic moments.

At 1122 hours 2. *Doppelpatrouille* under *Hptm* Wiesendanger took off.

At 1124 hours *Oblt* Heiniger set his J-324 on its belly in the direction of the Swissair Hangar.

The *Patrouille* Künzler/Schoch was with the B-17 "Blues in the Night" at the same time in the direction of Greifensee-Volketswil.

Oblt Künzler saw a P-51 approach the Dübendorf airfield from the direction of Zurich in a steep dive, but thought that it concerned *Hptm* Läderach with the interned P-51B.

In the middle of the attack on *Oblt* Heiniger, the B-24 "Lonesome Polecat" was in a landing approach to Dübendorf, where it touched down at 1126 hours.

After circling several times, the B-17 "Blues in the Night" landed at 1143 hours in Dübendorf. The final alarm sounded at 1200 hours.

After the last attack on *Oblt* Heiniger, both Mustangs flew in the direction of Kloten, and left Switzerland over Brugg and Frick at 1130-1133 hours.

The 2. *Doppelpatrouille* under *Hptm* Wiesendanger most likely saw the B-17 and the Messerschmitt escorts over Brüttisellen, but neither saw the Mustangs, nor was witness to the attack.

The *Patrouille* continued to fly in the direction of Frauenfeld, and were over the lakes by Nussbaumen when the radio message "follow in westward direction" arrived.

Hptm Wiesendanger and *Oblt* Siegfried thereupon flew in the direction of Olten without sighting either Mustang.

On the morning of September 5, 1944, 18 P-51s of the 339th FG under Lt Col Scruggs stood ready for the mission.

The takeoff in Fowlmere, UK, took place with 17 aircraft at 0838 hours. Northeast of Paris, over Crépyen-Valois, the unit ran into the bombers of the 3rd BD.

During this major offensive on the German industry centers, all Fighter Groups of the 8th AF were deployed in action. During the evening the U.S. fighters reported 27 shoot-downs, among these two by the 339th FG. The 339th FG did not suffer any losses in the mission, and landed in Fowlmere at 1445 hours.

It is uncertain if both P-51 B/Cs of the 503rd FS with Erikson and Ostrow arrived at the same time.

1st Lt Ostrow was accredited 1 ½, and 2nd Lt Erikson ½ shoot-downs. This could mean that during the attack Ostrow was involved with Treu as well as Heiniger, while Erickson was involved only in one attack.

The 8th Air Force lost a total of seven bombers during Mission 605, among them both aircraft from Dübendorf.

During the belly landing of J-324 the lower engine casing, together with the oil cooler, was torn off.

134

Hits from 12.7 mm bullets on Oblt Heiniger's aircraft. The P-51 B/C had at their command four machine guns.

The death of *Oblt* Paul Treu was especially tragic, because the Me 109 E of the Swiss *Flugwaffe* in Autumn 1944 still did not have armor protection in the cockpit, and the armored Me 109 Gs were several days previously barred from air service. Additionally, the radio equipment was classified at best as efficient.

The question arose regarding who had failed and had to be called in for an account. Major Dombrowski was entrusted as examining magistrate in the case.

It was essential to clarify why all messages of the Fl.B.M.D. were not fully conveyed. Of the six received messages at the A.W.Z. 5A, only two were fully passed on to the A.W.Z. of the *Kdo* Fl.B.M.D.

Radio contact to Messerschmitt was conveyed from the E.Z. in Bern over the transmitter "Emil" to Kleine Scheidegg.

A radio communication measuring flight by *Oblt* Heiniger on the morning of 5 September yielded negative results. The question was asked why the reserve transmitter in Dübendorf was not put into service.

Also, in the report of the *flab* there were contradictions. The *Schiessoffizier* of *Flab Det* 95 gave the time of the fire of the B-17 at 11:5 to 1127 hours, a time during which the bombers were already over Dübendorf.

The reproach was severe as to why *Fl Kp* 7 did not prepare their aircraft to be ready for shooting. An order that was valid since September 21, 1943, stated that munitions had to be supplied to the weapons.

In flight, to be prepared for firing the pilot had only to conduct a loading movement, or to turn the safety lock. *Hptm* Läderach ordered for the mission on 5 September "If an approaching aircraft is sighted: load all weapons, but not secured."

The affected pilots maintained, however, that before the particular mission there was no order given, while a pilot in the reserves confirmed the order from *Hptm* Läderach.

The fact was that the weapons in this mission were neither loaded nor released. In the aircraft of *Oblt* Treu, however, the FF-K were found released, while the MG 29 was not loaded (this meant that no bullet was in firing position).

During the emergency landing the fuselage was broken. The covering of the landing flap was probably tromped by the rescue team. Notice the outlet for the radio testing unit "TSF" and the oxygen line.

The heavily damaged underside of J-324. The fuselage was written off as a total loss. The repair of the wings and the motor cost 100,000 francs.

Examining magistrate Dombrowski came to the conclusion that no pilot could be blamed for "deliberate disobedience or non-compliance with the regulations," because the corresponding order of *Kpt Kdt* was not administered until after the mission, or did not reach the pilots in time.

It remains to be determined that the surprise attack of both Mustangs would have displayed the same effect if all weapons had been ready for firing. The problem was that the pilots could not be reached via radio.

The behavior of the American crew was likewise subject to criticism. In the foreground was why both Mustang pilots proceeded to attack although the nationality of the Messerschmitt and of the bomber crew was known.

In the B-17, Cpt Jaspers was the Command Pilot (*Kommandant* of the *Bombergruppe*). Jaspers, who controlled the radio equipment, would have sent the Mustang back to the border.

Because the navigator did not carry along any maps of Switzerland, he alleged to not have known that they were over Switzerland; navigators of the USAAF had to, in fact, carry along the maps that were necessary for the target approach.

Several members nevertheless had Swiss maps at their disposal; for example, the navigator of the B-24 "Lonsome Polecat," who also found Dübendorf without an escort.

It appeared odd that in the aircraft of the Group Commanders there were no corresponding map materials. There clearly were diverse ambiguities between the pilots 1st Lt Gallagher and Cpt Jaspers. While the navigator stated he would have also recognized the Swiss border without a map, and the pilot confirmed that they were in Swiss airspace, Jaspers was uncertain until landing in Dübendorf and acted nervously. He also made contradictory statements on whether he had recognized the Me 109 as a Swiss aircraft or not, as well as during radio contact with the Mustang.

The examining magistrate came to the conclusion that the 22 year old *Hauptmann* and Group Commander was not equal to the situation. Because both Mustang pilots could not be questioned, important questions remained open.

An accusation of "manslaughter by culpable negligence and bodily harm in terms of Art. 120 and Art. 124 of the MSTG" was dropped.

Along with the death of *Oblt* Treu and the wounding of *Oblt* Heiniger, civilians had also come to grief. Shots of a Mustang struck into the "Epileptische Anstalt Zürich," where two people were minorly injured.

The cockpit of J-324 with numerous bullet holes. Oblt Heiniger was not hit.

Aircraft of Fl Kp 8 with and without identification stripes in Thun. J-346 is equipped with an EW propeller.

In the evaluation of the diverse inconsistencies there were various opinions between the *Kdt* of *Fl Rgt 2 Oberst* Koschel and the *Kdt* of *FF Trp Oberstdiv* Rihner. While Koschel perceived the Americans to be primarily guilty, Rihner assessed the behavior of the B-17 crew in accordance with the circumstances as correct, and criticized moreso the attention of the Swiss.

Finally, because no one could be held directly responsible for anything, Major Dombrowski concluded the file on October 17, 1944.

When on 10 September a Mustang of the USAAF in the Pruntrut area again attacked two C 3603S of Fl Kp 11, measures were imposed to better mark the aircraft. It was obviously difficult for the U.S. pilots to differentiate the Swiss national emblems from the national identification cross of the German *Luftwaffe*. The Messerschmitts presented a problem, and also a C 3603 could be well mistaken for a Me 110.

The agreement of the bomber crew that was involved on 5 September had revealed that the Mustang pilots had many times questioned what kind of aircraft were in proximity to the bombers, because they had recognized these as Me 109s. Evidently, the bomber crew could also not clearly recognize the identity of the aircraft.

On 15 September the instructions followed to make all war and liaison aircraft better recognizable with red-white stripes. In the case that Switzerland would become involved in the war, it was planned to paint over the identifying stripes in the course of a night. All accredited *Militärattachés* in Switzerland were oriented on this measure.

Over 2,200 border violations were recorded until the end of 1944: 371 Allies, 158 of the Axis Powers, and 1,683 unknown. The "unknown" were ascribed to the mostly nightly overflights of the RAF.

With over 1,700 reports until the end of the war, an increase of border violations was to be recorded. Among these were merely 13 aircraft of the Axis Powers whose crews wanted to leave Switzerland.

A highlight was the 29 attacks within the Swiss border, among these the bomb releases on Zurich and Basel.

May 9 was declared "Victory Day" by the Allies, after already several days earlier the hostilities ceased in Europe successively on all fronts. On 12 May the staffs and units of the *Fliegertruppe* were released. Only the UeG remained on standby for emergency duties until the official end of active duty on August 20, 1945, in order to be able to take over possible neutrality protection assignments.

After the incidents of September 1944 the aircraft were provided with noticeable painting. Landing accident of J-367. Unterbach, October 27, 1944.

137

The B-24H (41-28994).
On September 12, 1944, the "Chatanooga Choo Choo" of the 455th BF was placed in the Schwarzsee area at 3,000 m altitude by a *Patrouille* of Fl Kp 9 under Oblt Häberlin. The pilots performed carefully. It was the first mission since the incident over Dübendorf on 5 September.

2nd Lt Bowling, a battle tested pilot with over 70 missions, did not react to the signal rockets of the Swiss *Jagdpatrouille*. In order to enforce the landing request, Oblt Häberlin shot a tracer in front of the aircraft. Bowling did not take a chance, and landed in Payerne at 14:08.

The B-24H (41-28994) with the *Mineur-Gruppe* of Kpl Philippe Clot, who was assigned with the destruction of military objects. Also, Payerne was the place of refuge for numerous U.S. bombers.

J-320 of Fl Kp 9 over the Alps.

Accident at takeoff in Interlaken on October 18, 1944. Due to a defect in the propeller adjustment, J-321 of Fl Kp 15 rolled into the Lütschine. The damage amounted to 100,000 francs.

Dübendorf in Winter 1944/45. The parking area on the airfield gradually became tight. Heavily damaged aircraft were often disassembled and brought to Kloten. After the war they were recycled as scrap metal.

Two Me 109Es of Fl Kp 8 with a neutrality paint scheme.

The pilots of Fl Kp 8 in 1944. From left: Oblt Schwarzenbach, Oblt Schärer, Lt Ernst, Oblt Brunner with hat, Oblt Dosch, Oblt Weiss, Oblt Dannecker, and Lt Eha.

A peaceful encounter: Me 109s of Fl St 8 escorted a B-17G-85-VE (44-8846) of the 305th BG. In 1947 the aircraft was on a surveying flight by order of the Federal Office of Topography. Hptm Max Brenneisen, surveying engineer of the Federal Office of Topography, was Kdt of Fl St 8.

Messerschmitt aircraft that did not belong in the inventory of the *Fliegertruppe*. The Me 262 arrived in Dübendorf on April 25, 1945. The aircraft underwent an extensive test, but was not flown.
The Me 110 G / 2Z+OP of NJG 6 landed in Dübendorf on March 15, 1944. Here the FuG 202 "Lichtenstein" was of special interest.
In the background is the Macchi C 205.
Dübendorf, April 1947.

A last Me 109 for the *Fliegertruppe*. J-399 was delivered in Altenrhein by Doflug on March 13, 1946.

140

Formation of Me 109Es.
It possibly involved one of the last flights of *Flieger-Staffel* 8 in 1949 (see color photographs on page 199).

Emergency landing due to propeller malfunction.
Oblt Friedrich Egli, Fl Kp 15, Avenches, June 27, 1946.
The aircraft was written off as a stockroom reserve.

A severe accident took place on October 2, 1946.
In the Raron region, two Me 109s of Fl Kp 9 collided during air combat practice.
In a right turn, J-344 grazed the cabin of J-339 with its propeller. Through the emerging imbalance the engine was dismantled, and smashed the cabin of J-339.
Oblt Kilchenmann and Oblt Vivian did not survive the accident.

The narrow undercarriage chassis was and remained the weak point of all Me 109s.
Presumably a Böe led to the downfall of the Kdt of Fl Rgt 4 in J-303.
Meiringen, February 9, 1948.

The Decommission of the Me 109 D and E

As of Summer 1940, the Me 109 Ds were only deployed as aircraft of the second line, and were not firmly assigned to any *Fl Kp*.
The first *Fliegereinheit*, which decommissioned the Me 109 E, was in Summer 1945 *Fl St* 21, followed by *Staffeln* 6, 9, and 15 in 1947. After the time on the Me 109G, *Fl St* 7 flew used the Me 109E until Summer 1947. *Fl St* 8 retired their long-serving Emil at the end of 1949.

In the last year of operation, approximately 560 hours were completed on the remaining Me 109D/Es on 953 flights (see flight statistics on page 235).

But the withdrawal of the old Messerschmitt Me 109 by no means indicated a stop forward for the *Fliegereinheiten*.

The successor, the D 3801, regarding performance, was classified at best as equal, and completely outdated regarding the technology of the aircraft construction. Not until the introduction of the Mustang and the Vampire did they have modern aircraft at their disposal.

After the withdrawal, the Me 109 D and Es were professionally dismantled at the *Betriebsgruppe* (Airbase Buochs).
Single *Baugruppe* (assembly groups) were further used in the Pilatus P-2.
Nothing is known of the manner of liquidation of the Me 109 G.

The tragic end of the Me 109: on the Forel flyer shooting range on Lake Neuchâtel, the cannibalized Me 109s were used as flyer targets. At this time the 8 cm rockets by Oerlikon were introduced—the effect of these weapons on ground targets was devastating. Forel, August 1950.

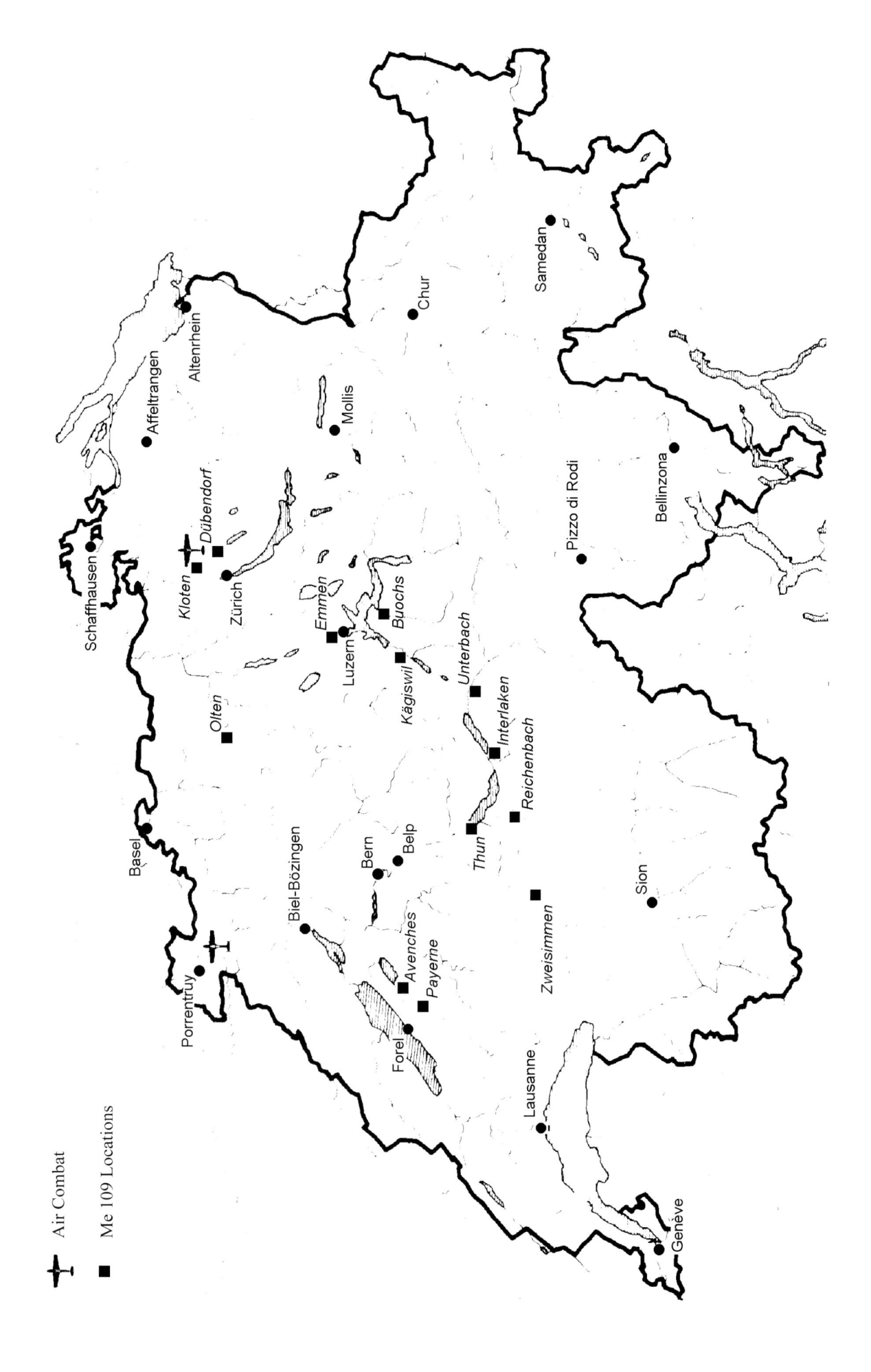

Outline Map / Events of Summer 1940-1949

- ✈ Air Combat
- ■ Me 109 Locations

143

Chapter 13:

The Me 109 F

The Emergency Landing in Bern-Belp

The invasion of the German *Wehrmacht* in Russia in Summer 1941 demanded the transfer of numerous *Luftwaffe* units to the East. On the Western front JG 2 "Richthofen" and J 26 "Schlageter" remained back. Each *Geschwader* had assigned in the *III. Gruppe* as *10. Staffel* Me 109 F-4 *Jagdbomber*.

During the last afternoon of July 25, 1942, *Fw* Villing and *Obgefr* Scharf took off from Le Bourget, Paris, for an overflight to Munich. Both young pilots belonged to the *Ergänzungs-Jagdgruppe Süd* of JG 5.

Fw Martin Villing planned a "visiting flight" at Stockach, by Lake Constance, his native city. Due to the poor weather, however, they could see neither the Rhine nor the Black Forest. When the blanket of clouds opened they saw an unfamiliar city below them. In a low level flight they tried to determine the name by the train station—it was Bern. Further eastward they caught sight of several gliders and an airfield. *Fw* Villing landed with his F-4/Z at 1930 hours in Bern-Belp. *Obgefr* Scharf assessed the runway as too short, and searched in the direction of Langenthal for a suitable landing spot. However, when the pilot light lit up, he decided to fly back to Belp, where he landed at 1945 hours.

Fw Villing's blue 10, a F-4/Z (Factory No. 7605), was constructed as PC+JY in the Wiener-Neustädter-Flugzeugwerke (WNF).

Henschel Flugmotorenbau in Kassel manufactured the DB 601 E with the Factory No. 34241. It flew in on December 20, 1941, in Wiener-Neustadt. The aircraft was assigned to 10.(*Jabo*)/JG 2, the *Jagdbomber* sorties against England.

The white 9 of *Obgefr* Scharf was a F-4 of the 10.(*Jabo*)/JG 26. The aircraft, Factory No. 7197 (NW+KU), was likewise built at WNF, and in August 1941 was flown in.

On November 8, 1941, *Uffz* Hofmann of 1. *Staffel* I./JG 26, with the Factory No. 7197, had to make an emergency landing by Hothem in Belgium. Hofmann was the *Rottenflieger* of the legendary *Staffelkapitän*, *Hptm* Josef Priller.

In the area of Dunkirk the engine stopped, but by pressing down it started and ran normally. A short time later the engine ceased again due to lack of fuel, which led to an emergency landing on a field. The damage amounted to 20 percent. The aircraft was subsequently brought into the Erla-Frontreparaturwerk VII in Antwerp. After the repairs on 21 October, an acceptance flight took place. The aircraft was later allotted to the III. *Gruppe* of JG 26 as *Jagdbomber*.

Both Me 109Fs in Bern-Belp.
In the foreground, Factory No. 7605 of 10. (Jabo)/JG 2. In the hangar, Factory No. 7197 of 10. (Jabo)/JG 26.
The aircraft still had their original markings.

The Me 109 F-4/Z, Factory No. 7605, of Fw Martin Villing.
The aircraft was built in the WNF in December 1941.

The Internment of Martin Villing and Heinz Scharf

After being questioned by the intelligence service in Hotel "Chaumont" by Neuchâtel, both pilots were interned. Villing had doubts that they would be accused of desertion in Germany. In fact, the *Gestapo* interrogated the family members. The German *Militärattaché*, *Oberst* Grippe, however, could allay doubts. Furthermore, they received a monthly service pay of 93 francs for *Fw* Villing, and 30 francs for *Obgefr* Scharf. The German *Kolonie* in Switzerland paid for civilian clothes. Both could freely move, and also came together with interned Allied pilots. School children even demanded autographs.

Later the Italian Alfredo Porta consorted with them. He had landed in Emmen with a Fiat G.50 on 21 September.

On 24 August a Mosquito PR IV of the RAF landed in Bern-Belp. The Allies, also like the Axis Powers, wanted to at least have their pilots back. In a diplomatic manner an agreement could be reached. When both British arrived at the English embassy in Madrid after a trip over Vichy-France, Villing, Scharf, and Porta could also return back to their homelands on December 21, 1942.

Heinz Scharf lost his life in 1944 in a scramble, and Alfredo Porta was shot down in air combat.

Martin Villing survived the war. He was deployed with JG 5 "*Eismeergeschwader*" under the hapless Major Erler in Norway. Sections of JG 5 were responsible for the security of the battleship "Tirpitz." When on November 12, 1944, the RAF sunk the ship by Tromsö, the culprit was sought in Major Erler, and he was court-martialed. Martin Villing ended the war with 21 air victories.

The Me 109 in the Swiss *Flugwaffe*

For the time being both Me 109 Fs remained in Bern-Belp. The aircraft underwent extensive inspection, whereby the cold start device attracted special interest. After starting the engine, during 20 to 30 seconds a certain amount of gasoline would be supplied to the lubricant.

The *Kdt* of *FF Trp* asked the KTA if such a device could be utilized for the Hispano-Suiza engines of the D 3800/01.

At the beginning of 1943 both aircraft were brought into F+W in a dismantled condition. A diagram of the main migration by head mechanic Köbi Urech could mean that around 26 February test flights had taken place.

Because the aircraft were not at the disposal of the troop, in the flight statistics of the *Fliegertruppe* no flights on a Me 109 were listed.

Martin Villing (left), Alfredo Porta (a Swiss organizer), and Heinz Scharf in the Neuchatel Jura.

By Swiss standards both aircraft had few operating hours:

Aircraft	Cell h	Engine h
No. 7197	40.04	38.58
No. 7605	43.01	31

Between the RLM and the EMD negotiations took place on the purchase of both aircraft, each for 300,000 francs.

Furthermore, an exchange for two German Me 109 E-7s was suggested. The E-7 was a further development of the E-3/E-4, with the possibility of installing an additional 300-liter tank, and an ETC 50 or ETC 500 bomb carrier. The aircraft with Factory Nos. 3198 and 4973 were planned for the exchange. However, the deal was never achieved.

As all interned aircraft that were not bought by Switzerland, the Me 109 Fs were also not utilized during the war in the troop. Only in Summer 1946 were the numbers J-715 and J-716 issued. Unfortunately, no photograph is known that proves that the matriculation in fact was applied. However, the aircraft never came to the troop, and were dismantled in Emmen at the beginning of 1947.

The white 9 of Obgfr Heinz Scharf with the emblem of the 10. *Jabo Staffel* of JG 26.

The blue 10 of Fw Villing.
On the engine cowl, the emblem of the 10. *Jabo Staffel* of JG 2.

Factory No. 7605 had round wheel cases, while No. 7197 had angular ones at their disposal.
Notice the camouflage painting.

The Me 109 F-4 presented a great advancement, as opposed to the Swiss Me 109 E-3.
The experts identify the "F" as the best-balanced variant of all Me 109s.
The armament consisted of two 7.92 mm MG 17s and a central 2 cm cannon MG 151/20s.
Factory No. 7197.

The armor of the Me 109 F was clearly influential.
The armor plates on the canopy that were installed in the Swiss Me 109 E in 1945 had similar proportions.
Factory No. 7605.

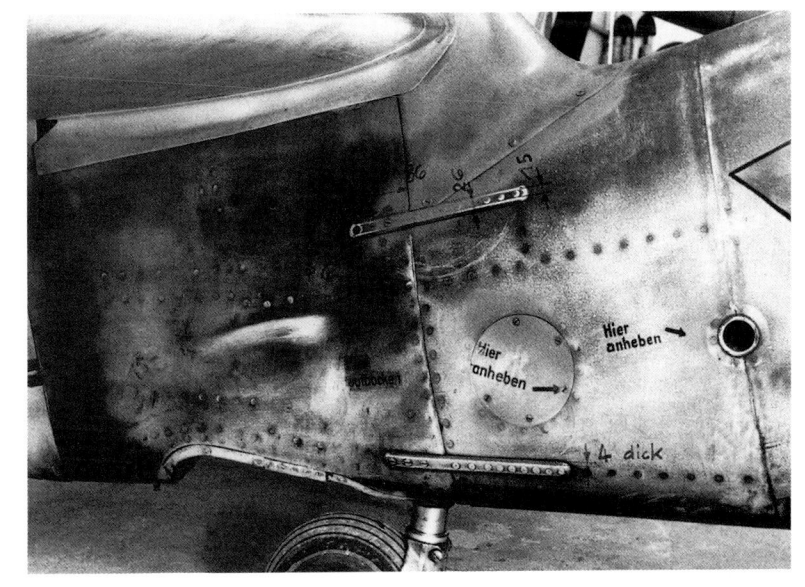

The lightweight construction of Prof. Messerschmitt also showed its weaknesses with the Me 109 F. Through the cantilevered tailplane strong forces appeared with frame 9 that led to the break of the tail body.
Through the riveting of reinforcements remedial action was taken.
Later the structure of the rear body was reinforced.
Factory No. 7197.

Factory No. 7605 in F+W Emmen, April 1943
The camouflage paint was changed.

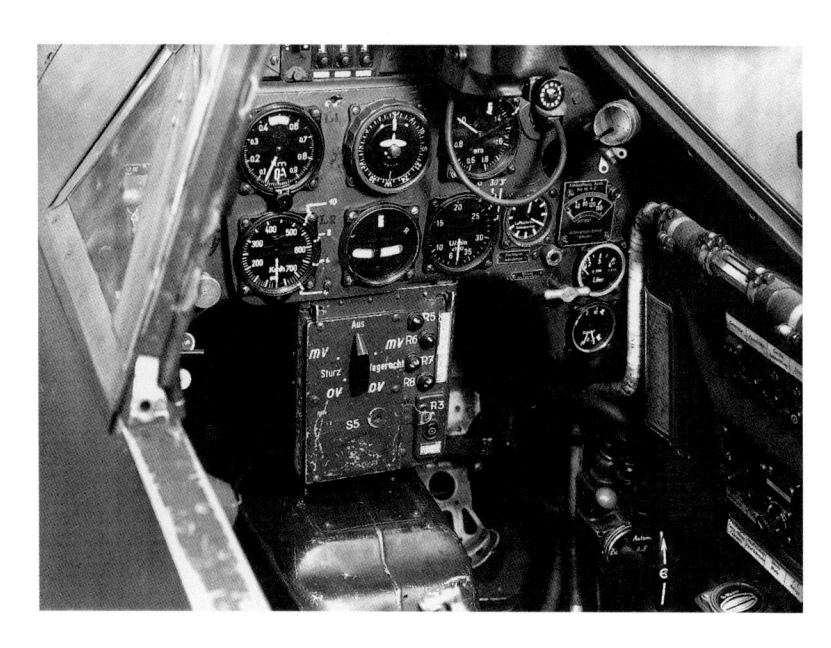

The cockpit of Factory No. 7605
In the middle, the control unit for the bomb assembly. Underneath the covering for the MG 151/20.
The *Waffenwarte* had to climb headfirst into the cabin in order to load the weapon.

A picture out of focus, however, the most meaningful documentation of an ME 109 F with a Swiss National Emblem. An aircraft number was not appointed. In 1946 the number J-715 was assigned. It is not documented if it was also painted on. Factory No. 7605, F+W Emmen, July 1943.

Chapter 14:

The Procurement of the Me 109 G-6

1944, and Still No Modern Aircraft

In Spring 1944, a delivery of 30 training aircraft (Arado Ar 96 Bs and 40 radar Würzburg Ds) was negotiated with Germany in a counter deal to short wave AS 59 and AS 60 transmitters from Switzerland.

At the same time, one speculated if a procurement of aircraft from England was possible.

Under the condition that if the political situation in Europe would change within a conceivable time, and that a rail transport or an overflight from England into Switzerland would be feasible, one first and foremost planned for the Spitfire IX. In second place was the Typhoon IB.

It was a question of principle regarding the engines. The English had aviation gas at their disposal with more than 100 OZ. The fuel that Switzerland had available had merely 93 OZ, which would have led to a decrease in engine performance.

In a writing dated April 5, 1944, the head of the KTA asked the *Militärattaché* in England to clarify if a delivery of circa 50 aircraft would be possible.

The matter was delayed for over one year. On April 19, 1945, the *Kdt* of *FF Trp* proposed a resignation of the by now outdated Spitfire model Mk IX.

Interlude in Samedan

At the end of March 1944 one could orientate oneself for the first time in Switzerland on a Me 109 G-6.

On 29 March, at 1110 hours a brand new aircraft with Factory No. 162764 and the mark of origin RU+OZ landed in Samedan. In the logbook was the name Me 109 G-6 / J Bo Rei, which indicated a *Jagdbomber* with increased reach.

The 23 year old *Oberfähnrich* Lothar Hirtes, by order of the 2.(Süd)/FlÜ.G.l., was to flyover the aircraft from the Erding airfield to Osoppo, on the Southern Front. According to the files Hirtes belonged to 4./JG 5.

During the worst weather Hirtes lost his orientation while crossing over the Alps, and landed on a mountain airfield unfamiliar to him against the landing direction. The *Kdo Armeeflugpark* under *Oberst* Högger determined that circa 100 liters of fuel were available.

On March 18, 1944, RU+OZ flew into Regensburg, and a flight authorization was issued until 21 September.

On 27 March RU+OZ, at the *Luftflotte* 2 airfield in Erding, was ready to be given over to the troop with 1.45 operating hours.

The introduction of the Me 109 G brought Switzerland a short but turbulent time with high performance aircraft.

The single known photograph of RU+OZ.
On March 29, 1944, the aircraft landed in Samedan. At the end of August RU+OZ was issued the number J-713. As most interned aircraft, it was integrated into the Flugwaffe only after the end of the war.

J-713 (ex RU+OZ) was in Buochs as of Summer 1944. On November 8, 1945, it was flown over to Altenrhein, then prepared at Doflug, and afterwards given to the troop.

RU+OZ was militarily fully equipped. The FuG 25a, subject to secrecy, was provided with an explosive charge for self destruction. It is not known why Hirtes did not destroy the device.

On 1 April the aircraft was flown over from Samedan to Dübendorf. In Summer 1944 RU+OZ was disassembled in F+W Emmen. The *Gruppe* Hug in Buochs received the assignment on August 23, 1944, to make the aircraft fit for service as a J-713, analogous to the Me 109 G introduced in the meantime. At the same time, the bargaining over the price of the aircraft began. The Germans, represented by Rittmeister von Eggen, demanded 500,000 francs, half of which in free foreign exchange. However, in Switzerland one was of the opinion that either way the aircraft of the German *Luftwaffe* was no longer available, and would be adequate at a price of 250,000 francs. Thereupon, Berlin still demanded 400,000 francs. Because this sum was also perceived as too high, one forewent a purchase in October.

Until the end of the war the aircraft remained unused in Buochs, and was flown over to Altenrhein on November 8, 1945. On 3 December J-713 was test flown, and declared ready for service.

The Mysterious *Nachtjäger*

On April 28, 1944, at 2 AM German *Nachtjäger* of NJG 5 in the pursuit of bombers of the RAF arrived in Swiss airspace. In the Dübendorf area they were caught and pursued by flab searchlights. *Oblt* Johnen was thereby so irritated that he was forced to land on the Dübendorf military airfield with his Me 110 (C9+EN).

A second Me 110 was piloted by *Fähnrich* Mutke. However, he was able to escape, and continued his flight in the direction of France. Over Nancy the crew had to escape the aircraft with parachutes due to lack of gasoline. The Me 110 crashed into a residential house at the end of the airfield, whereby four people were killed. Guido Mutke came again one year later, this time by his own choice, on a Me 262 to Dübendorf. (The incident cited above was taken from the investigation report on *Fähnrich* Mutke from the year 1945. In 1997 Herr Mutke confirmed its accuracy to the author).

The Me 110 G-4 of *Oblt* Johnen was equipped with the SN-2 "Lichtenstein," and the aircraft weapon equipment "Schräge Musik." Johnen was an experienced *Nachtjäger*, and was able to register 34 shooting downs until the end of the war. The aerial gunner, *Fw* Mahle, was decisively involved in the installation of the *Schrägbewaffnung* in the Me 110. The radio operator, Lt Kampath, had a file with codes, tables, and maps that he would have never been permitted to take with him on a sortie.

The Me 110 G-4 of Oblt Johnen.
The incident about the aircraft brought Switzerland to the edge of a military conflict with the Greater German Reich.
Notice the up-firing armament is removed.

The Obertraubling airbase at Regensburg after the end of the war. The Messerschmitt-Werk II was also here, in which the Me 109 G was assembled.
The facilities were badly destroyed above all in February 1944 during the "Big Week."

The landing of this aircraft—interesting for Switzerland, and also the Allied secret services—incited diplomatic chaos between Berlin and Bern. The German Government immediately recalled the aircraft. For legal reasons of neutrality this was, however, rejected. After hard negotiations—Germany had even considered a *Kommando* action and bombardment of the Dübendorf airfield—it was decided to detonate the aircraft in the presence of the German military in Dübendorf. In return one held out the prospect to Switzerland of a delivery of twelve Me 109 G-6s and Würzburg radar.

The detonation of the Me 110 took place during the late evening of 18 May. Already one day later the EMD requested the purchase of twelve Me 109 Gs. At the *Bundesrat* session on 20 May the deal was concluded.

The price of 500,000 francs per aircraft was observed fit. The necessary 6 million francs were taken from the account "*Aktivdienst Materialbeschaffung*," and paid in free foreign exchange.

The representative from the Reich Main Security Office, *SS Hauptsturmführer* Rittmeister von Eggen, preferred to obtain the sum in gold. Eggen was to have directly given Göring two checks at a value of 3 million francs, who placed them "immediately in his breast pocket."

The fact is that the *Luftattaché* in Bern, *Oberst* Gripp, did not know to whom the checks had gone. In a writing from October he maintained that a check of 3 million francs was directly given over to Rittmeister von Eggen personally during the delivery of the first six aircraft in Dübendorf.

Thus, for the Swiss *Flugwaffe* a short but turbulent period with Messerschmitt aircraft, accompanied by rumors, speculation, and anecdotes, began that has held to this day.

A Special Topic: The Production of the Me 109 G

In the Swiss *Flugwaffe* the Me 109 G did not have a good reputation. While the flight performance was never doubted, however, the technical flaws, especially those of the engine, were a central problem.

The rumor is still persistently held today that the aircraft were built for merely 20 operating hours. However, there are neither technical documents, nor any documents that prove this. In contrast, documents exist on the inspection and testing of the aircraft that prove that these operations were diligently carried out.

The only known photograph of a Me 109 G at its delivery in Dübendorf.
Because only few knew about this deal, the aircraft caused a stir with the German route marking.
Factory No. 163 251? RQ+BO was matriculated with J-706.

It may be correct that many aircraft on the Front reached few operating hours before they were lost in action or an accident. It is accurate that through the simplified operations, extremely bad working conditions, and the use of unskilled forced laborers, the quality standard of the pre-war period was no longer achieved.

With the manufacture of approximately 44,000 aircraft of all types in Germany in 1944, the highest production number during the war was reached. This impressive number, however, was only possibly through the inconsiderate use of forced laborers. The production of the Messerschmitt aircraft belonged to the darkest chapters in the history of military aviation.

In literature little or only suggestive material can be found on the Me 109, but what is relevant in connection with the criticism of the Me 109G is dealt more closely with here.

The assembly groups and accessory parts for the Me 109 were not only produced in the Messerschmitt main factories, but also in private major enterprises right up to small enterprises, as well as in numerous concentration camp field warehouses.

The influence of the SS in the armaments industry became greater and greater in the course of the war. It was the assignment of this "*Schutz-Staffel*" to procure the necessary personnel for the industry. Women and men from occupied regions, concentration camp prisoners, and prisoners of war were forcibly recruited. With the most brutal methods these people were forced to work, and had a corresponding failure rate due to accidents, malnourishment, sickness, or death.

These people, degraded to slaves, were not only manpower who were cheap and without rights, but also an extremely lucrative business for the SS. They earned millions with the "hiring" of prisoners.

The Messerschmitt company availed abundantly in this way. Prof Messerschmitt was also blamed for this after the war. How many forced laborers in the main factory, the branch factories, and the numerous suppliers were employed will probably never be determined—at any rate it had to have been tens of thousands.

The Me 109Gs for Switzerland that came from the Regensburg factory were finally assembled in Vilseck or Hagelstadt.

As a result of the Allied air attacks, the manufacture of single assembly groups was largely decentralized. Large components, such as wings and fuselages, were built in the Flossenbürg (Oberpfalz) and Mauthausen/Gusen (Austria) concentration camps. The assembly, equipment, and testing furthermore took place in the Regensburg area.

In the Flossenbürg concentration camp prisoners were employed in the SS *Schatten-Firma* "*Deutsche Erd- und Steinwerke GmbH*" (DEST) under the perception of "prisoners as workforce."

As of 1943 the mining in Flossenbürg faded from the spotlight. The Regensburg Messerschmitt GmbH established contact with the SS at the highest offices of party and ministries. For Messerschmitt it concerned being able to employ prisoners from the Flossenbürg and Mauthausen concentration camps for aircraft production. In February 1943 component part production began in Flossenbürg, initially with approximately 200 prisoners. Additionally, a section of the Messerschmitt production in the factory buildings of DEST was shifted to the Flossenbürg quarry area. After corresponding factory buildings were rebuilt, the sheet metal formation work to complete the Me 109 began.

Messerschmitt delivered the raw material, and made available devices and machines. The foremen, Meisters, and engineers commandeered by Messerschmitt normally acted correctly toward the prisoners. The works management could even receive an increase in food rations for the forced laborers. Additionally, there were premiums for particularly good performance. However, this should not belie the fact that in Flossenbürg thousands of prisoners lost their lives: as of August 1943 an average 800 prisoners, as of December 1943 already 1500, and as of March 1944 2,200 prisoners were employed in the aircraft production. If it was only initially engines and tail assemblies, practically the entire aircraft was built in Flossenbürg. For reasons of secrecy the production had the cover name "*Kommando* 2004."

The Gusen field warehouse of the Mauthausen concentration camp was another establishment of DEST. These, also called "*Georgen-Mühle*" warehouses, were likewise changed from a quarry to an armament industry.

As of Spring 1943 wings and fuselages for the Me 109 were built. The prisoners who worked for Messerschmitt also worked in better conditions here than those in the quarry. However, this changed when at the end of 1943 the *Sonderkommando* Kammler envisaged Gusen as an underground armament center. Under the cover name "*Bergkristall*," in 1944 the field warehouse Gusen II emerged. In the underground facilities above all the Me 262 was produced and finally assembled.

The working and warehouse conditions had to have been murderous: in the three warehouses of Gusen approximately 37,000 people lost their lives until the end of the war.

It is obvious that those conscripted were not motivated to build aircraft for their tormentors, or in the case of prisoners of war for their enemies.

Acts of sabotage, such as sand in the oil tank, filed hydraulic lines, coated screw connections, and slashed gasoline tanks had to be dealt with.

Still today it is discussed among the experts if one deliberately delivered defective or used aircraft to Switzerland. At any rate, the total fabrication number point to new aircraft that came from a running series.

To what extent one knows of the incidents in Germany in Switzerland gave and gives rise to discussions to this day.

There is no indication if representatives of the Swiss *Flugwaffe* were in Regensburg at this time. Only a name of a head mechanic was verbally given to the author who was to have been in the Messerschmitt factories. Whether this is true, and if so, it is not known which factory he was in.

The Me 109 G-6 for the Swiss *Flugwaffe*

During the socalled "Big Week" (February 20-25, 1944), during which the 8th and 15th USAAF attacked German armament centers, while the Messerschmitt factories in Regensburg were also hardly hit.

Due to the constant threatening air danger the factory airfields by Regensburg had to be cleared again and again. For testing the aircraft additionally were brought to alternative airfields, such as Puchof or Obertraubling. According to the situation, this took place in overflights, or by means of road transportation with dismantled wings.

The inspection of the aircraft was supervised by *Hptm* Obermeier from the BAL (*Bau-Aufsicht der Luftwaffe*).

The test flying lasted normally 20 to 25 minutes, and the acceptance flight circa 30 minutes. Afterwards the aircraft were placed in the surrounding forests until delivery.

The flight files of RU+OZ are maintained. From this it is evident that the acceptance and control of the aircraft were carried out at least diligently during this time.

The *Überführungsstaffel* 3.(*Süd*) Fl.Ü.G.l. under *Hptm* Weissmüller (cousin of "Tarzan" Jonny Weissmüller) was responsible for the transport to Switzerland. The *Gruppenkommandeur* was *Oberstlt* Fachner in Neubiberg by Munich.

The first six aircraft from the April production were supposedly tested in Vilseck, and on 20 May flown over from Regensburg over Neubiberg to Dübendorf. The second six were possibly tested in Puchof, and came from the May production. The overflight to Dübendorf took place on 23 May.

The aircraft initially caused a stir in Dübendorf. Only few knew about the deal with the Germans. The sudden appearance of German aircraft in Dübendorf was seen as a betrayal to Switzerland.

The pilots who had to fly in civilian clothes were not able to enjoy any Swiss hospitality: just after the delivery of the aircraft in Dübendorf and a short refresher they were again brought over the border in a bus with covered windows.

The aircraft were immediately provided with Swiss national emblems and matriculation, and came into air service around May 26, 1944.

Factory No.	Route Marking	Overflight	Aircraft No.
163 112	ST+RB	05-20-44	(J-701)
163 320	ST+VP	05-20-44	(J-702)
163 243	RQ+BG	05-20-44	(J-703)
163 245	RQ+BI	05-20-44	(J-704)
163 248	RQ+BL	05-20-44	(J-705)
163 251	RQ+BO	05-20-44	(J-706)
163 804	NF+FE	05-23-44	(J-707)
163 806	NF+FG	05-23-44	(J-708)
163 808	NF+FI	05-23-44	(J-709)
163 814	NF+FO	05-23-44	(J-710)
163 815	NF+FP	05-23-44	(J-711)
163 816	NF+FQ	05-23-44	(J-712)

The delivery of the Me 109 G

Factory No. 163245 / RQ+BI after the release to the Swiss Flugwaffe in Dübendorf. Notice the directional loop and the absence of the antenna for the FuG 25a.

Parade lineup with eleven Gustav in Interlaken, June 1944.
J-705, which was disabled due to a deformation of the wing truss.

Munitions Procurement

In contrast to the Me 109D and E, the Gustav was delivered with armament and munitions, and to a large extent corresponded to those of the German *Luftwaffe* regarding equipment. To what extent checking devices, special instruments, and ground equipment were procured is today no longer able to be determined.

The original demand for an aircraft with an engine cannon was first fulfilled with the MG 151/20.

In the munitions box of the twelve aircraft there was a total of 836 cartridges with armor piercing projectiles. A further delivery of 1,000 tank projectiles and 2,000 incendiary grenades from Germany occurred in July 1944.

The Bührle company in Oerlikon was assigned the manufacture of the corresponding munitions shortly after the delivery. Already in mid-August tank, steel, and practice grenades could be delivered. The 13 mm munitions for the MG 131 presented a problem.

Aside from the munitions delivered with the aircraft, at the end of July and beginning of October further deliveries followed directly from Germany.

The procurement of munitions from Germany, however, was not easily possible due to the state of war. The Bührle company was therefore requested for the fabrication of 13 mm munitions.

The electrical caps of the projectiles were subject to nondisclosure. Because no technical documents could be obtained, Bührle had to refuse the order for the time being. Thereupon, the KTA checked the installation of the MG 29 in the Me 109G. The installation would have been technically possible, however, had there been considerable reconstruction work involved.

In order to optimize the supply of the munitions and the discharge of the shells and elements, one would have had to configure the weapons spacing analogous to the Me 109E.

The installation of two MGs of rifle caliber would also have meant a step backwards in weapons technology.

In the meantime, from Germany 200,000 caps were released for shipment to Switzerland. When in October a further delivery of 13 mm munitions from Germany arrived, the installation of the MG 29 was left aside.

Furthermore, the Bührle company also succeeded in manufacturing the caps without records. In Spring 1945 the production of munitions in Switzerland could be taken up.

Antennas and Radio Equipment

With the FuG 16Z one had a high performance UKW transmitting and receiving device in the wave range of 7.1 to 7.8 meters at their disposal. A rebuilt FuG 16 likewise served as a temporary ground station.

Although the directional loop PR 16 was applied to the aircraft, no homing device or radio direction finder was installed. Later the directional loop was removed.

In the single known photograph that shows a Me 109 G-6 during delivery to Switzerland, the antenna for the FuG 25a is visible.

The cockpit of Me 109 G.
Notice the Revi C/12b, and the flare gun Walther 4002.

The IFF device FuG 25a "Erstling" worked in connection with Freya-Würzburg or Gemse stations, and was subject to secrecy. Switzerland had none of the corresponding equipment at their disposal. There is no indication if the FuG 25a was installed in the Swiss Me 109G.

Erla Hood and Augmented Rudder Assembly
J-701 to J-706 aircraft were equipped with the customary rudder assembly and the strutted canopy.

J-707 to J-712 had a socalled Erla or Galland hood, the extended rudder assembly, and tail plane made of wood.

The improved engine performance of the DB 605 with the corresponding increased torque demanded an increase in lateral stability. This was attained with the increase of the rudder assembly in which the end cap of the tail assembly and rudder were replaced with higher end caps. Until the withdrawal from service, all aircraft were retrofitted with the high rudder assembly.

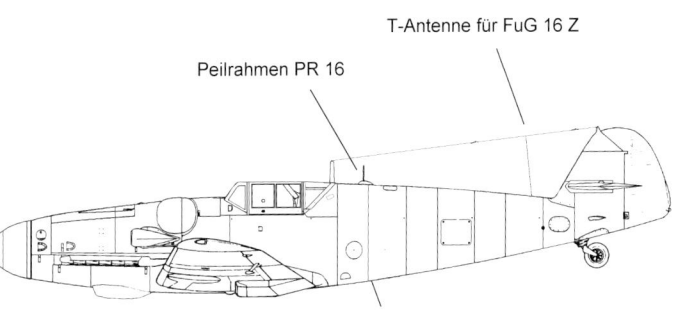

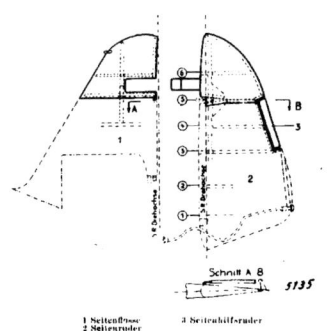

With the Me 109G the *Flugwaffe* for the first time had at their disposal an aircraft with satisfactory armor.
J-701 to J-706 aircraft were originally provided with a strutted canopy.
The visibility for the pilots was not optimal with this hood.

J-707 to J-712 aircraft were equipped with an Erla hood that provided better visibility.
Notice the antenna mast is found on this aircraft at frame 2, and not as usual on the cabin frame.

Trop Tanks and Bombs

In a document of the KTA from December 1944, it was revealed that at Messerschmitt twelve trop 300 l tanks were ordered. In this context an armament with bombs was also tested.

The Me 109G belonged to the Usage Class H and the Stress Group 4 or 5 based on stability regarding equipment, and the reinforced condition of the wings.

During takeoff and landing with a full trop tank or bombs, the manufacturer advised caution in its instructions, because with this external weight the breaking point was reached. External weight, as was noted, should be dropped before landing.

The Me 109G corresponded to the H5 Class with an acceptable flight weight of 3,200 kg. With a MW assembly and an additional 300 l tank, the aircraft corresponded to the H4 Class with an acceptable weight of 3,600 kg.

There is no indication whether the additional tanks were ever delivered. There is also no document or photograph on which the tanks or the corresponding rack are visible.

Unforeseen Gain

In mid-June 1944 the procurement of 20 further Me 109Gs was considered. One wanted to equip a second *Staffel* with 12 aircraft, plus 4 standbys, and the already existing unit with four standby aircraft. In addition to this were spare engines, spare materials, and munitions. However, the deal never was achieved.

On 19 July a North American P-51B of the 8th USAAF landed in Ems. As of August the aircraft was tested in Dübendorf, and subsequently utilized for air combat training. For the first time one could directly compare the newest technology from America and Germany.

An unexpected accrual of two Me 109G-6s, at least as a source of spare parts, arrived on August 20, 1944.

Flg Nehrenheim from 5./JG 77 attempted an emergency landing on the troop drill ground Bern-Beundenfeld. From an eastwardly direction he prepared for landing on the Beundenfeld. The yellow 3 with Factory No. 163956 was caught at the end of the ground in a trench and hole for hand grenades, whereby the aircraft overturned. The pilot was minorly injured on the face.

One minute later, at 0800 hours *Fw* Tanck in yellow 12, Factory No. 163226, likewise landed on the Beundenfeld. The aircraft collided with a tank mockup, brushed against several trees, and crashed down a street. The pilot was uninjured, but the aircraft was heavily damaged.

The pilots were on a ferry flight to Northern Italy and lost their orientation.

Thereby, one may ask the question how it was still possible in Summer 1944 that two foreign aircraft could fly completely uninterrupted into the federal capital....

The yellow 3 of 5./JG 77 in Bern-Beundenfeld on August 20, 1944.
The pilot was lucky; he was only slightly injured.

A procurement of additional tanks was certainly considered but never actualized.
Whether the rack and tank of Factory No. 163956 were used is not known.

Factory No. 163226 collided on the Beundenfeld with a tank mockup. The pilot remained uninjured.

Both Me 109G-6s that landed on the Beundenfeld served the Swiss *Flugwaffe* as welcomed spare parts donors.

The Me 109G-14 with the marking of III./JG 3 slid into a drain on December 17, 1944, at Affeltrangen.
In 1946 the aircraft J-714 was taken into the stock of the *Fliegertruppe*.

Factory No. 462818 was provided with a MW 50 assembly. Notice the triangular identification for the methanol/water mixture.

Numerous Me 109G-6s and G-14s, which had no pressurized cabin, had at their disposal, however, an engine cowling that was common with the Me 109G-5 with pressurized cabin (recognizable at the air inlet, and the protrusion for the compressor under the buckle of the MG 131).
This cowling section was found again on J-704.

On December 17, 1944, a German Me 109G-14 with the marking of III/JG.3 had to make an emergency landing by Affeltrangen.

Fw Siegfried Henning was on a ferry flight from Erfurt-Bindersleben into the Southern German region when he was attacked by a Mustang of the USAAF. In a low level flight he succeeded in escaping his pursuer, whereby he deviated from his course and lost his way in Switzerland. At 1612 hours he attempted an emergency landing with retracted landing gear on an open field. Thereby, a barbed wire was caught on the counter balance of the right aileron. Thus, the aircraft turned and landed in a drain. *Fw* Henning remained uninjured and was interned. In 1946 the aircraft was stored in the inventory of the *Flugwaffe* as J-714.

The "*Sonder-Kraftstoffanlage* MW 50" attracted special interest. In a report from the *Armeeflugpark* on December 30, 1944, that was classified as secret, it is mentioned that one will more closely check the equipment. However, there is no indication if the equipment was utilized in J-714 or in another aircraft.

One Last Interest for the Me 109: the Me 109 G-10/AS
In January 1945 the representative of the Messerschmitt factories of the KTA assured that "without further issues" 24 or even 48 further Me 109Gs could be delivered.

The entire cost for an aircraft was quoted at 920,000 francs. From this 450,000 francs were planned for the aircraft itself, 150,000 francs for spare parts, and 320,000 for munitions. From this sum 300,000 francs was to be paid in foreign exchange. Meanwhile, the *Fliegertruppe* no longer showed any interest in these aircraft.

On February 1, 1945, the KTA, in accordance with the *Kdt* of the *FF Trp*, proposed a rejection of further Me 109Gs from Germany. On one hand, the experiences with the twelve available aircraft were unsatisfactory. On the other hand, it was uncertain what quality aircraft would be delivered.

Factory No. of the aircraft raises several questions: Although shown differently on the rudder assembly, Factory No. 462818 is listed in the board files. This block number points to a G-14 from Erla.

158

No growth for the Swiss *Fliegertruppe*.
On March 26, 1945, at 1330 hours at Werthenstein/Lucerne, a German Me 109G rammed into the ground.
Fähnrich Herbert Jehring had to leave his aircraft behind with the parachute due to engine problems. He landed uninjured circa 1 km east of the location of impact of his aircraft.

It was further argued that, despite the noticeable identification stripes the Messerschmitt aircraft presented a certain security risk for the pilots. At this point it was no longer maintained politically or opportune to procure aircraft from Germany.

In a writing dated February 16, 1945, the EMD officially renounced the purchase of further Me 109Gs.

However, with this the problem of spare part procurement, spare engines, and munitions was not solved.

A last reference on the procurement of Messerschmitt aircraft is located in a German (classified secret) telex from February 9, 1945.

A sender in code "S GENST 6 ABT KURFUERST" asked the German *Luftattaché* in Bern for a statement: At Messerschmitt the KTA proposed a Me 109G-10 with DB 605 AS for the purpose of a demonstration.

Evidently, this proposal did not proceed in an official manner, and it was not elaborated on.

In the turmoil of the last weeks of war a serious deal with the Third Reich would no longer have been possible.

...nterlaken airfield in August 1944, with a view in the direction of Wilderswil and the foothills of the Schynige Platte.
The airfield was the base of Fl Kp 7.
In the foreground, J-712 after a landing accident.

A Gustav of Fl Kp 7 in Summer 1944.
It was not clarified for what reason the painting was applied. At the end of the '30s Fl Kp 7 had a shark as the Kp emblem. In Summer 1940 a more peaceful trout served as a symbol. In 1944 one probably wanted to accentuate a certain aggressiveness, and painted at least one aircraft in the present manner.

Below: List of DB 605 engines used in Switzerland.

The problems with the Me 109G were primarily found in the motor. Like airframes, in Germany at this time engines were also produced by forced laborers.

Aircraft Marking	Manufacturer	Engine No.
HSS	Niedersächsische Motorenwerke Braunschweig-Querum	00 202 165
		00 203 253
		00 203 676
		00 203 790
HSG	W. Dieck Apparatebauanstalt, Aussig-Türmiz	00 703 975
		00 703 871
HSR	Henschel Flugmotorenbau, Kassel	01 104 624
		01 102 872
		01 102 929
		01 103 896
		01 104 217
		01 104 652
Spare Engines (Manufacturer HSG?)		00 709 782
		00 709 867
		00 709 883
		00 709 928
		00 709 985

Engines from interned Aircraft
Factory No. 162 764 / RU+OZ (J-713   01 102 434)
Factory No. 462 818 / <2+I (J-714   01 101 442)
Factory No. 163 956 / Yellow 3   00 705 781
Factory No. 163 226 / Yellow 12   21074

160

Chapter 15:
The Me 109G in Action

The Disappointed General

For the first time in Switzerland, aircraft with good armor and reliable radio equipment was available. The flight performance was convincing, and moreover, the Gustav was praised as a stable shooting platform.

The aircraft were initially stationed in Dübendorf, were flown by test pilots and pilots from UeG, and maintained by the ground personnel of the DMP. To what extent they were supported by corresponding technical personnel at Messerschmitt is today no longer able to be determined.

On June 9, 1944, the pilots of *Fl Kp* 7 enlisted for retraining from the Me 109E on the "Gustav" in Dübendorf. During the afternoon the first flights were carried out, whereby predominantly takeoff and landing maneuvers were practiced.

For 10 June the overflight to Interlaken (the home base of *Fl Kp* 7) was planned. The *Oberbefehlshaber* of the army wanted to see these "super aircraft" for himself. General Guisan waited in vain. As only two aircraft were ready to fly, the practice was canceled.

In the following weeks the aircraft were flown from Interlaken and Payerne without noteworthy problems. The use of the weapons took place on the Forel flyer shooting range at the end of June.

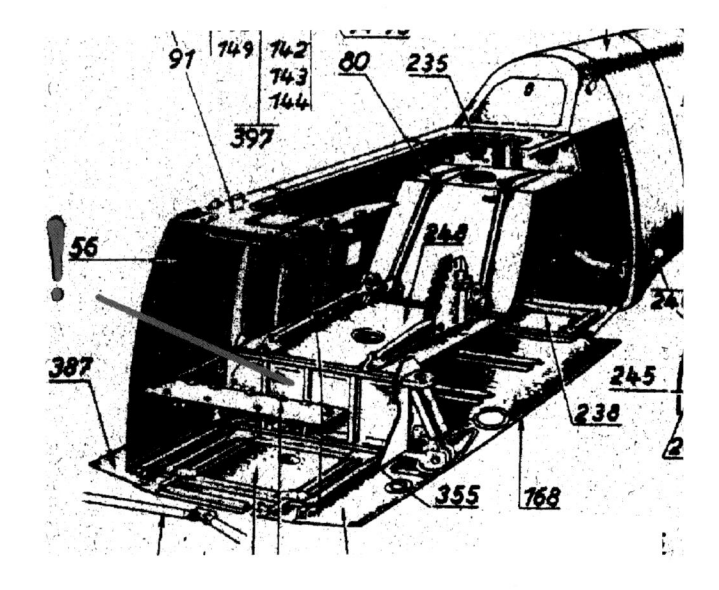

Unknown Sounds

In several aircraft a cracking sound was heard occasionally during taxiing on the ground and during flying. Investigations revealed that this was caused by an elastic deformation of the wing truss.

On J-703 and J-710 one found a bulge in the midfield of the wing truss with the mark "B." This was presumably labeled by a German control station as *Beule* (bump).

The wing truss of J-705 displayed such a stark deformation that the aircraft had to be disabled.

If the deformation was caused by overstress in flight or during a poor landing is unable to be determined.

The question of repairs, and the possible reinforcement of the wing truss on aircraft was immediately settled with the F+W Emmen. The further course of this matter is not known.

On some Me 109Gs the ring trusses were deformed.

J-708 in Thun, Summer 1944. Notice the yellow oil cooler covering and the directional loop PR 16. The antenna of the FuG 25a is not fixed on.

During landing training in Interlaken on August 19, 1944, the right landing gear collapsed on the aircraft of Oblt Treu. The antenna mast for the FuG 16 Z was fixed on the cabin frame. The directional loop PR 16 is in the distance.

On August 4, 1944, Oblt Feldmann had to make an emergency landing with J-706 in Interlaken. The reason was engine failure at 8.55 operating hours. This was the first in a long row of incidents with the Messerschmitt Me 109G.

The Engine Problems

On 4 August *Oblt* Feldmann had to make an emergency landing in Interlaken in J-706 due to engine failure. The technical investigation revealed that the supercharger was eroded due to the lack of lubricant. The cause was an assembly defect at the manufacturing company. A spacer sleeve, which at the same time served as the oil supply ring, was not installed.

The engine no. 00 207 390 was brought to S.L.M. in Winterthur for maintenance. One tried to blame the manufacturer for the cost of 20,000 francs.

J-704 had to make an emergency landing on 26 August. A bearing of the crankshaft was defective, which led to irreparable damage. The engine no. 01 104 217 had to be written off with only 6.42 operating hours, and from then on served as a spare parts donor.

After this incident, on 2 September all aircraft were suspended from air service. All of the engines were disassembled, and underwent inspection by the S.L.M.

The inspection and revision reports were sobering: none of the twelve engines came off without complaint (see table below). Furthermore, the following defects were determined:

- In all of the engines the screws were loose, improperly or not at all secured.
- At the rocker levers to the valves the bases were not cut rectangular.
- The supercharger drive was poorly balanced.
- The bearing sleeves of the water and oil pumps had too much slackness.

Because the Me 109G was suspended from air service, *Flieger Kp* 7 was equipped with nine Me 109Es on 4 September. One day later *Oblt* Treu was fatally shot down in his unarmored Me 109 E by a Mustang of the USAAF (see page 132).

The first Me 109Gs—among these J-701 and J-710—came back to the troop in the course of October 1944, while others were decommissioned until mid-January 1945.

Fluttering Stabilizers

Aluminum alloys were not available in abundance in Germany, as well. Therefore, one looked for suitable substitution materials.

Tailplanes and rudders were henceforth increasingly made of wood, which also reduced the costs by approximately 20 percent.

The tail assemblies of J-707 to J-712 displayed manufacturing defects: when exceeding a speed of 700 km/hr a dangerous vibration occurred. The highest speed—that is, the reading on the speedometer—had to be reduced to 650 km/hr.

Engine No.	Aircraft No.	Hours	Defect
00 202 165	J-705	22.22	Tear in cylinder liner No. 1
00 203 253	J-709	18.34	Various missing and loose screws
00 203 676	J-701	24.23	Loose and knocked out bearing sleeves
00 203 790	J-706	8.35	Oil supply ring missing
00 703 871	J-710	20.22	Various loose screws and riveting
00 703 975	J-712	17.06	Broken through and covered screws
01 102 872	J-707	18.41	Lockwasher on large end bearing No. 2 too long
01 102 929	J-708	21.11	Poor valve seat
01 103 896	J-711	17.00	At bevel gear drive to oil supply pump spline missing
01 104 217	J-704	6.42	Crankshaft bearing defect, engine written off
01 104 624	J-702	17.32	Various assembly defects and broken components
01 104 652	J-703	18.28	Spark plug thread nipple loose, cylinder block wall burned through

The cause of engine failure in J-706 was an erosion of the supercharger drive. The oil supply ring (pos A) was missing, whereby the bearing was not greased.

The completely destroyed supercharger drive of DB 605A1 No. 00 203 790. For the damage of 20,000 francs one attempted to hold the manufacturer liable.

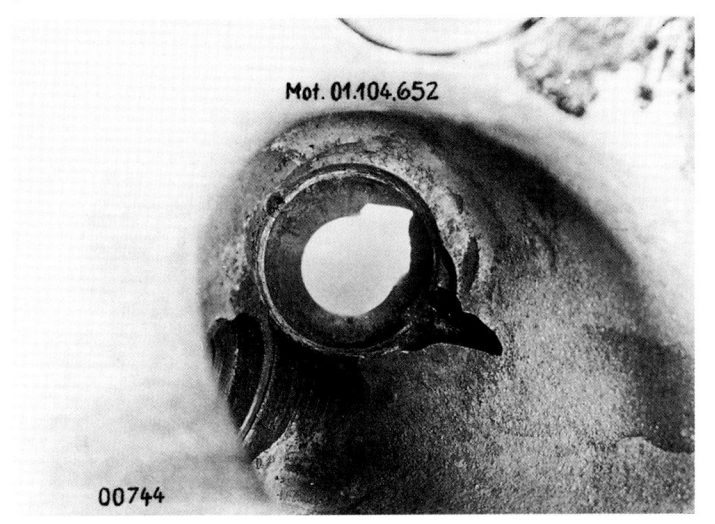

In the cylinder 4R of the engine of J-703 the lockwashers to the terminal end of the M2 spark plug were missing. As a result the spark plug could loosen; the result was the burning through of the cylinder block wall. The engine 01 104 652 had at this point in time 18.28 operating hours.

On 26 August J-704 had to make an emergency landing as a result of engine failure. The cause was a bearing defect on the crankshaft. The engine 01 104 217 had to be written off with merely 6.42 operating hours. In the photograph an eroded bearing pin.

On the supercharger of the engine No. 00 203 253, installed in J-709, various clamping bolts were missing. The bore of the bearing was thereby knocked out.

J-711 with identification stripes. Notice the additional inscription OZ 93. The reason for this was that the Me 109G could not run on K fuel. Interlaken, June 1945

A reproduction of wooden tail assemblies was not possible in Switzerland. Substitutions had to be obtained from Germany—no easy endeavor at the end of 1944.

The exchange or the reinforcement of the defective tail assemblies took place possibly at the beginning of 1945 in F+W Emmen. Presumably, at the same time J-701 to J-706 were also retrofit with the increased rudder assembly.

Until the end of 1944, on 703 flights and with the twelve Me 109Gs a total of 205 hours were flown.

Around January 20, 1945, the Me 109G *Flotte* was again suspended from air service: On the engine of J-711 a locking pin on the spark plug nipple had fallen out, which resulted in an inspection of all engines.

Fl Kp 7 once again had to fall back on the Me 109E. Not until April during the TK of the *Rgt* 2 were the Gustavs again allocated.

On January 3, 1945, a mechanic in Interlaken carried out a static test with J-704. At start-up the engine sprang into full blast, and the aircraft slid over the brake shoes.

After circa 75 meters the mechanic jumped off, and the pilotless aircraft broke through the fence, crossed over the street, and remained on the shore embankment of the Lütschine. J-704 was completely destroyed. The investigation revealed that the throttle valve stood open at start-up at 35 instead of 15 degrees.

The wreckage was brought to Buochs and newly constructed. Thereby, the wings were exchanged. Whether the wings were new, or if they came from both aircraft from Bern-Beundenfled is not known.

The damage amounted to 200,000 francs. The repairs lasted several months. On September 10, 1945, the aircraft was again ready for flying. J-704 presented itself with a completely new camouflage paint and an increased vertical stabilizer.

On October 22, 1945, *Htpm* Läderach lost the canopy in the Basel area because it was not correctly locked. J-704 was thereupon provided with a Erla hood.

Oblt Hofer was in J-705 over Martigny on 22 November when suddenly the warning light lit up. Approximately 5 km before Sion the engine failed. The landing approach proceeded as normal, but while pulling through the aircraft lost height. The landing gear caved in, and J-705 slid circa 100 meters on its belly. The investigation revealed that in the pressure-tight gasoline tank a layer had come off and sealed the suction tube.

On April 13, 1946, at 1342 hours *Hptm* Wiesendanger, *Kdt* of *Fl Staffel* 7, was approaching Buochs. Due to a breech of the driving claw the propeller adjustment abruptly changed, and the aircraft crashed to the ground at a steep angle of descent. *Hptm* Wiesendanger thereby suffered a head and back injury. J-710 was completely destroyed.

Also, there were other accidents due to pilot errors: On May 25, 1946, in Altenrhein a pilot forgot to extend the landing gear. J-701 crashed, which resulted in damage of 20,000 francs.

A Gustav of Fl Kp 7 at compensation. Interlaken, Winter 1944/45.

On January 3, 1945, in Interlaken J-704 swerved during static testing. The pilotless aircraft came to a stand on the shore of the Lütschine. Note: The aircraft had no identification stripes.

The forward section of the fuselage, together with the engine were broken off during the accident. J-704 was brought to Buochs and newly constructed.

After the repairs the repaired J-704 was presented in new camouflage paint. Notice the expanded rudder assembly and the strutted canopy. Buochs, September 10, 1945.

165

Emergency landing of Oblt Hofer in Sion on November 22, 1945. A removal of the rubber layer in the gasoline tank interrupted the fuel supply.

In November 1945 J-705 was still provided with the socalled neutrality paint. Notice the expanded rudder assembly in the original paint of the German *Luftwaffe*.

The *Betriebsgruppe* Buochs was the branch for the Me 109. In the hangar the Me 109E and G with identification stripes. To the far left a Me 109E in priming, coat and with an armor plate on the canopy. Mechanic Fritz Heller shot this photograph of the assembly hangar in Summer 1945.

During a landing failure on the grass runway J-707 wrecked a latrine. Emmen, June 26, 1946.

The end of J-710 in Buochs on April 13, 1946.
Hptm Fritz Wiesendanger, named "Bschütti-Fritz," did his nickname credit when he had to make an emergency landing on the airfield cultivated with fresh manure.

Due to a break of the driving claw to the propeller adjustment the aircraft crashed to the ground in a steep gliding angle.
Hptm Wiesendanger thereby suffered a back injury, and J-710 was a total loss.

The accident triggered renewed discussion on the operational security of the Me 109G.
Unpopular aviation orders could be dismissed with quote "Jä, wüssed sie Herr Oberscht, mit em Gustav cha mer das Nöd mache."

The propeller together with the transmission was knocked off during impact.
At the time of the accident J-710 had 106.11 operating hours.
Notice the exit opening for the signal pistol. On the windshield the drying body and fuel spray line for windshield cleaning are visible.

The Me 109G did not have any landing gear alert available. Oblt Kind's undoing on May 25, 1946, in Altenrhein.

J-701 had at its disposal an altered camouflage paint and an increased rudder assembly; however, it still had the old, strutted canopy.

For once it was not the narrow undercarriage that caused an accident—the pilot merely forgot to lower it.

After a faulty start in Buochs on June 14, 1946, J-706 was damaged.
There is no indication whether the aircraft was repaired.
Officially J-706 was written off on September 8, 1947.

Right photograph
J-706 was flown over to Interlaken to Fl St 7 by Oblt Mirault, pilot of UeG-Staffel III.
The damage amounted to 40,000 francs.
Notice the lower four assembly holes for the sand filter and the maintenance instructions.

The cockpit of J-710 after the accident of Hptm Wiesendanger.
Buochs, April 13, 1946.

The Forgotten J-713

The flight of *Oblt* Zweiacker of *Fl St* 7 ended tragically.

He took off on May 29, 1946, at 1110 hours in Emmen in the direction of Tessin. Over the Gotthardpass he had to turn around due to a blanket of clouds. The aircraft then circled over Erstfeld and flew via Surenpass to Engelberg. Over the Sustenpass the aircraft was heard for the last time. *Oblt* Zweiacker was declared missing from then on. Not until September 4, 1953, did mountain climbers find the remains of J-713 on Pizzo di Rodi. According to the location of the debris the approach took place from Airolo, the northern direction. The cause of the accident could not be determined.

J-713 (ex RU+OZ) in Altenrhein, December 1945.
Because Switzerland and the German Reich could not reach a purchasing price of the aircraft during the war, it was allotted to the *Fliegertruppe* just after the end of the war.

The debris of J-713 lay widely scattered in the cliffs. Access to the scene of the accident is allowed only a few weeks during the year due to the snow. It was pure coincidence that the wreckage was discovered.

Oblt Zweiacker and his aircraft were found just in 1953 by mountain climbers on the Pizzo di Rodi (TI).
The cause of the accident is unknown.

Due to engine malfunction Hptm Widmer had to make an emergency landing in Interlaken. In doing so he rammed two parked C 3603s.

The End of the Me 109G

The end of the Gustav commenced in November 1946.

On 13 November at 1551 hours DMP pilot *Hptm* Widmer took off in J-704 in Interlaken for an inspection flight. A short time later an engine failure occurred due to a connection rod break, and the cabin filled with smoke. Over the Rugen *Hptm* Widmer threw off the cockpit canopy and attempted an emergency landing. Hindered by the strong smoke, he taxied over the street Interlaken-Bönigen in the direction of Hangar 2 and rammed two parked C 3603s.

J-704 suffered a total loss; the repairs of the C 3603s cost 125,000 francs.

A few days later on 20 November *Oblt* Hofer from *Fl St* 7 in Payerne had to make an emergency landing in Payerne in a J-707 due to engine failure—the shaft of the supercharger drive was broken.

Afterward all Me 109Gs were again suspended from air service. All DB 605s were disassembled and underwent inspection at the S.L.M.

On February 27, 1947, the status report was presented: all engines exhibited serious construction, material, and assembly defects.

For maintenance per engine circa 90,000 francs was estimated, always provided that one could procure replacement parts at all.

One attempted to come into contact with the Svenska Flygmotor company, because in Sweden the DB 605 was still in operation.

However, a newer engine ready for installation cost 191,000 francs, which was out of the question for reasons of credit.

At the same time one was in contact with the Fiat agency in Geneva. At Fiat, in Turin, brand new DB 605s were available. However, it was feared that these licensed engines would exhibit the same defects like those in Germany, and a purchase was foregone.

In view of the minor number of only eleven aircraft, and the prospect of procuring P-51D Mustangs, it was decided to decommission the Me 109G.

The proposal to write off the aircraft was presented by *Kdt FF Trp* on July 31, 1947, and six days later by the KTA to the Swiss military department in Bern.

Several months later one could procure 130 P-51D Mustangs from the excess stock of the U.S. Air Force. A Mustang cost only $4,000; at this time this was approximately 17,000 francs. The price for a DB 605 engine bore no relation to 191,000 francs.

Fl St 7 flew the Me 109G until Summer 1946. For the training course in 1947, the well tried Me 109Es were again used. The "*7ner*" retrained in 1948 on the D 3801—an aircraft that was to some extent reliable, but was generations behind regarding performance.

Until the end of 1946, 1,054 flights for a total of 491 hours were completed. There is no indication if in 1947 the Gustav was still flown. As of Summer 1947 the Me 109Gs were officially decommissioned.

J-704 was the last Gustav that was lost in an accident. The cause was a connection rod break on engine No. 00 203 790. Notice the engine cowl of Me 109 G-14, Factory No. 462818.

171

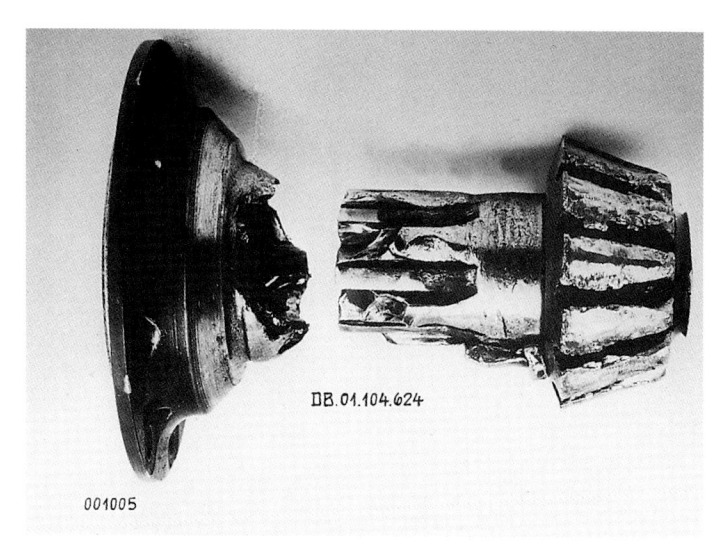

After a renewed emergency landing due to engine malfunction on November 10, 1946, in Payerne the Me 109G were definitively banned from aviation service. The engine 01 104 624 (J-707) had 90.13 operating hours. In the photograph the broken supercharger drive shaft.

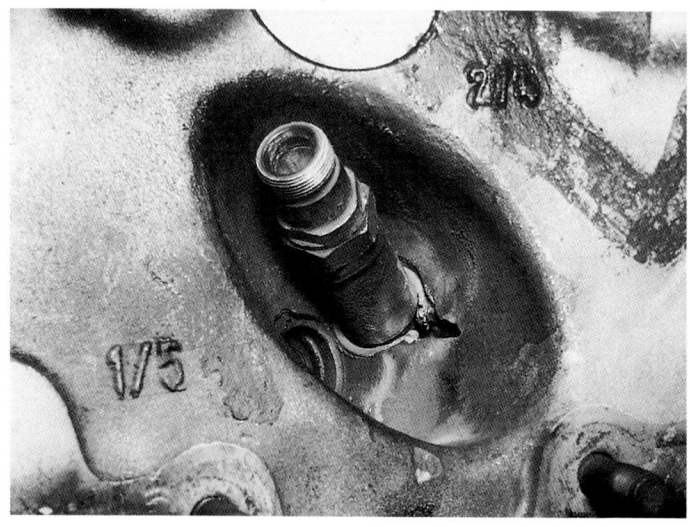

In all engines leaky and burned through spark plug nipples were to be found. A postproduction was not possible due to the marginal wall thickness of the cylinders.

On the engine 00 203 676 (J-702), after 117.19 operating hours various compositions of the crankshaft bearing were cracked.

In January-February 1947, in the SLM, all DB 605s underwent inspection. On 27 February the status report was submitted:

- Poor storage for the supercharger drive in the device carrier: defect in construction
- Leaky spark plug nipple on the cylinders: defect in construction
- Crankshaft bearing with cracked composition: defect in material and casting
- Crankshafts with breaks: defect in material
- Control shaft with breaks: defect in material
- Gearwheel on the propeller shaft with breaks: defect in material
- Inside connection rod with breaks: defect in material
- Ball bearings and roller bearings with too little compression: inadequate Seeger circlip ring security
- Pressure tube torn
- Burnt exhaust valves
- Breaks of screws
- Defect supercharger shaft fuses
- Poorly arranged casing, crankshafts and cylinders
- Castings with cast seams (risk of breakage)
- Poor sustaining control shaft bearings
- Poor fitting stud bolts

Breaks in the bearing eye of the inside connection rod No. 1. Engine No. 00 708 883 (J-708) at 91.26 operating hours.

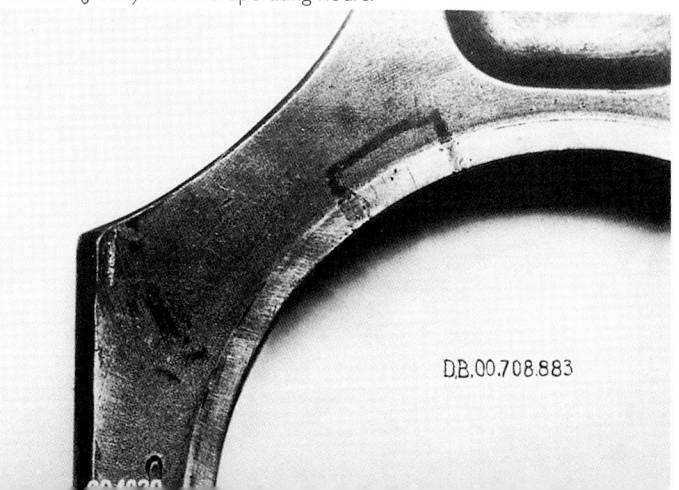

Break in the large internal gear of the DB 605 of J-711 at 113.46 operating hours.

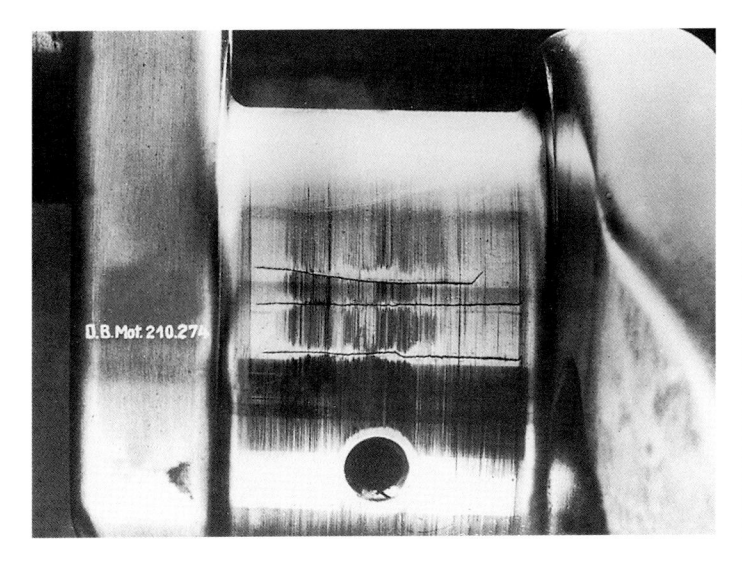

Pivot No. 5 on the ball bearing of the spare engine No. 210.274 showed three cracks. A break could have led to an emergency landing.

Foreign substances in the oil led to traces of corrosion on the pistons. Pistons 11 L, Engine No. 01 102 929 (J-703) at 136.11 operating hours.

Broken off bolts (Pos 4 o'clock) on the supercharger flange of J-701. Bolts were often covered during assembly. Acts of sabotage of this kind were easily carried out because they were hardly discovered.

In the engine No. 00 709 867, the locking pins were missing on an inside connection rod. The S.L.M. was blamed for this defective assembly.

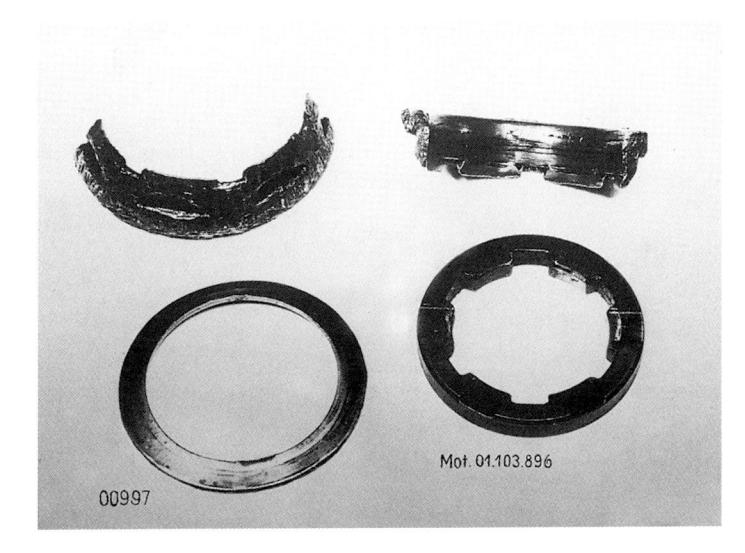

Broken union nut to the drive shaft of the supercharger on a spare engine. Due to defects of the supercharger, approximately three Me 109Ds, five Me 109Es, and three Me 109Gs had to make an emergency landing.

173

At the end of 1946 flight service operations with the Gustav were abandoned. There is no indication whether the aircraft were flown until the official withdrawal in Autumn 1947, respectively Spring 1948. In the photograph a VDM/9-12087 propeller with the No. 9533, mounted on J-713.

The Me 109G – Better than its Reputation

Without a doubt the Gustav was better than its reputation. The problems of the Me 109G were predominantly located in the engines. The airframe, deformed wing trusses, and the fluttering of the tailplane were the single noteworthy complaints.

At least in the early days the aircraft were gladly flown by the pilots. The flight performance was completely satisfying. Moreover, the Gustav was praised as a stabile shooting platform. This was displayed above all during the shooting approaches on the notorious Axalp.

Oblt Robert Heiniger completed 168 flights on the Gustav. Thereby, he flew all aircraft besides J-708 and J-713.

In the total 105 flying hours, a single defect appeared that had nothing to do with the aircraft itself. On July 7, 1944, the cannon on J-710 jammed during a shooting approach in Forel.

Due to the emergency landings and the various engine defects that were determined during inspections, however, doubt emerged of the operational safety of the aircraft. According to the *Staffel-Chronik,* the *7ner* did not mourn the Gustav.

The Me 109G was deployed in numerous air forces. The countries were, in contrast to Switzerland, in some way or another allied with Germany. This was significant for logistics. In this context, the use of the Me 109 in the Finnish Air Force was interesting.

The Finnish procured circa 160 Messerschmitt Me 109Gs of diverse variants. The last aircraft were decommissioned in 1954. Despite intensive war operations, numerous Me 109Gs reached over 300 operating hours.

Finally, it may be maintained that the twelve Me 109Gs of the Swiss *Flugwaffe* represented the single efficient aircraft at the end of active duty.

A possible successor for the Me 109—the Hawker Sea Fury. In the background is the Me 109G of Fl St 7. Interlaken, July 1946.

174

Chapter 16:

Camouflage and Markings

The Camouflage

The procurement of the Me 109 came at a time in which the type of camouflage paint was still not clearly defined. The paint was also not an issue during the purchase of the Me 109.

At this time the German aircraft were maintained with a "2 color visual cover" in RLM 70 (black-green) and RLM 71 (dark green). A single color paint in RLM 70 was the exception. The aircraft underside was always RLM 65 (light blue). The ten Me 109Ds for Switzerland were delivered in a single color paint RLM 70. The Me 109E of the first series had a camouflage pattern seldom used in the German *Luftwaffe*. This consisted of RLM 70/71, however, not in the typical camouflage pattern with straight, but rather curved lines. The number of aircraft with this pattern is not documented, but it had to be J-311 to J-340.

The underside of all Me 109D/Es was always kept in RLM 65.

Regarding camouflage of the Swiss Me 109, there are no documents available. Possibly such were never drawn up. The two-colored camouflage pattern can be understood only by means of photographs. While the right side of the aircraft was sufficiently documented with photographs, the left side is never clearly depicted. Only with documents on the camouflage pattern of German Me 109s can the color gradient of the Swiss aircraft be approximately reconstructed. Furthermore, the camouflage pattern presented several differences on single aircraft.

However, the camouflage pattern remained only a short time. In the course of the installation work for the MG 29, the outer casing was changed. The entire forward section of the fuselage was sprayed over with KW 2 (RLM 70).

The rear fuselage was presumably adjusted in April 1940 during the changing of the identification marking.

During this work the camouflage pattern was merely oversprayed; the demarcation of the old camouflage paint remained visible. On the wings, horizontal stabilizers, and vertical stabilizers the camouflage pattern was not removed.

The Me 109E of the 2. Series (J-341 to J-390) had the camouflage RLM 70/65 at delivery. The Swiss recreations J-391 to J-399 were maintained in KW 2 and KW 1.

As numerous photographs demonstrate, there existed no clear instructions of the paintwork. Several pictures of aircraft of the 2. Series display a camouflage pattern on the wings. It is uncertain if these originally or subsequently were applied. These possibly emerged during repair work with a changing of the wings.

The colors for the Me 108 and the Me 109 D/E were initially obtained from Germany.

As of Summer 1939 the shades were mixed in Switzerland, whereby they were also renamed. Compare RLM / K+W / DMP:

Deutsche *Luftwaffe*	K+W	DMP
RLM 70 Greenish-Black	KW 2 Dark Green	Art. No. 602 342 Dark Green
RLM 71 Dark Green	KW 3 Light Green	
RLM 65 Light Blue	KW 1 Blue-Gray	Art. No. 602 344 Light Blue

In Switzerland the RLM marking was not used.

In order to avoid misunderstandings due to the various definitions of shades, in this publication K+W colors are named.

By mixing the colors, in the course of the years the hues did not remain consistent.

The shade RLM 65/KW 1 was subject to a stark change. This was also apparent in the German *Luftwaffe*. Likewise, with KW 2 dark green, color shifts were determined that had nothing to do with the aging of the color.

The shade KW 3 was removed in 1946 from the "Instructions for Paintwork."

The Me 109D was delivered ex factory in schema RLM 70/65. Emergency landing in Ellikon ZH, January 5, 1939.

The Me 109Es of the first series were kept in schema RLM 70/71/65. The gloss level was semi-gloss, the national emblem and matriculation glossy. Dübendorf, Summer 1939.

Photograph below:
J-316 during revision work in the Malerei Buochs. The original camouflage paint is still visible.

The installation of the MG 29 demanded reconstruction work on the covering hood. Afterwards, the forward section of the fuselage was newly sprayed with KW 2. However, the camouflage schema on the rear fuselage remained. Emergency landing of J-327 in Dübendorf, February 29, 1940.

During the application of the new identification in April 1940, the camouflage schema on the fuselage was repainted on all aircraft.

The camouflage pattern on the wings; the horizontal and vertical stabilizer for the time being remained. Aircraft of the second series with a camouflage pattern on the wings were the exception. The taxiing accident in Payerne, February 11, 1944.

Factory No. 163245 (RQ+BI) after the application of the Swiss national emblem, Dübendorf, May 1944.

With the takeover of the Me 109G in May 1944 one was confronted with a new type of camouflage.

The aircraft were maintained in the then common shades of gray RLM 74/75/76. The fuselage side was patterned with RLM 02.

Presumably two aircraft were provided with a grounded hood. The possible shade (yellow, RLM 02, or RLM 66) cannot be properly determined.

In no Swiss document are there statements on the camouflage paint of the Me 109G. During conversations with aircraft painters from this time there is still only little to be learned. Obviously, the new shades in the painting workshops were mixed by discretion. Because the position of the German national emblem did not correspond to those of Switzerland, corrections had to be made.

It is uncertain if the overspraying of each area occurred with available colors of the K+W or with mixed shades of gray.

During revision work, as of Summer 1945, during which the identification stripes were removed from the fuselage, the camouflage on the fuselage side was correspondingly adjusted with light specks. Each aircraft was individually sprayed without exact guidelines. It is assumed that shades mixed by hand were used. According to the memory of a former aircraft painter, for the camouflage specks on the fuselage side KW1 blue-gray with white was lightened.

In September 1945 J-704 was provided with an uncommon camouflage paint. The green corresponded to the common KW 2 dark green, while the brown was modeled approximately after the German RLM 81 brown violet.

An access flap of J-704 is still available. A visual comparison with FS 595a (1979) shows a shade between brown 30099 and 30118. The underside was KW 1 blue-gray.

J-713, and also possibly J-714, were provided with a similar camouflage pattern; the shades KW 2 and KW 1 are documented. With the light shades it presumably involves KW 3.

There is no photograph known of J-714.

The original deep blue KW 1 or RLM 65 had in the course of time come very near to the shade RLM 76 light blue.

J-713 in Altenrhein, end of 1945.

The Me 109G in the post-war period was provided with an individually different camouflage schema. The colors involved the mixed hues RLM 74/75/65. However, there is no evidence of this, merely statements from the then involved parties. False start in Buochs, June 14, 1946.

Swiss military aircraft were provided with a round national emblem on the wings from 1938 until January 1940.

The rudders were originally only halfway painted in red. Notice the Kp number on the rudder assembly.

National Emblem and Markings

The aircraft number originally consisted of three numbers 50 cm in height. In 1938 the prefix "J" for *Jäger* at 40 cm high was introduced. The numbers were not positioned in the same manner on the Me 109D and Me 109E. The reasons for this are unknown.

The national emblem on the wings since 1938 were in the form of a roundel with a Swiss cross. To increase the recognizibility for the ground troops, in January 1940 a red band with a Swiss cross was applied to the wing underside. At the same time the entire area of the rudder was painted red.

A striking change took place in April 1940. Numerous aircraft of the Swiss *Flugwaffe* of the same model were flown by warring states, such as Germany or France. The recognizability of the Swiss national emblem was in many cases difficult. Additionally, on the fuselage side a red band with a Swiss cross was applied. The aircraft number was minimized to 25 cm high.

In order to prevent accidents with running propellers, in 1940 a white safety marking was added on the propeller blades.

The aircraft remained in this form until the hectic days in September 1944. The experiences from abroad and maneuver practice in Switzerland displayed that the ground troop was not able to recognize the nationality of aircraft until the last phase of an attack.

In July 1944 diverse special markings were tested to increase the recognizability of the aircraft. The socalled invasion stripes of the Allied Air Forces in Normandy probably also had an influence on these painting tests.

After the incidents with aircraft of the USAAF from September 5 and 10, 1944, measures were enforced. All Front aircraft were provided with red-white identification stripes. The engine cowlings received white paint.

In order to make easier the recognition of the nationality of an aircraft for the ground troops, in January 1940 the red band with the Swiss cross on the wings' undersides was again introduced. At the same time the entire rudder was painted red.

179

The identification of Swiss aircraft in the air was problematic, especially because the Flugwaffe flew German and French models. In April 1940 a red band with a Swiss cross was applied to the fuselage side and the Flz Number was reduced by 25 cm.

Until all aircraft were repainted both forms of the new identification had to be observed. Fl Kp 21, Dübendorf, April 1940.

In the style of the Allied invasion stripes, in Switzerland a similar special paint was tested. However, these kinds of identification stripes were never introduced. Interlaken, presumably July/August 1944.

180

After the incidents with fighter aircraft of the USAAF, in mid-September red-white identification stripes were introduced. This paint, which became well known as neutrality stripes, was maintained until after the end of the war.

The application of the identification stripes on hundreds of aircraft lasted for some time. Both variants were seen at the same time.

The identification stripes were applied in *Fl Rgt* 2 and 4 by the troops. Aircraft that were not in troop service were painted by the *Armeeflugpark* or by the industry.

In case Switzerland became involved in the war, it was planned to paint over the identification stripes over the course of one night. This noticeable painting was removed as of Summer 1945 during a part or total revision of the aircraft. For reasons of aviation security the identification markings on the wings were kept. The position of the aircraft numbers remained unchanged.

Me 109F – Open Question
Both Me 109Fs were maintained in the standard paint according to L.Dv. 521/1 (RLM 74/75/76). Furthermore, the factory no. 7605 was sprayed over with a dark shade.

In photographs from Spring 1943 in F+W Emmen, the factory no. 7605 is clearly seen in the shades KW 2 and KW 1. Unfortunately, there is no known photograph that depicts the aircraft in its entirety. Swiss crosses were applied in the then common form. An aircraft number was certainly not applied at this time. On one hand, it was not common to matriculate interned and unpurchased aircraft; on the other hand, the numbers J-715 and J-716 were not allocated until June 1946.

However, the aircraft was never received by the troop.

Until a photograph is available that provides an explanation, the question remains open if ever a Messerschmitt Me 109F existed in full Swiss decoration.

An order on September 15, 1944, states: in the case of acts of war with Switzerland the identification stripes are to be immediately painted over. Notice the lower applied Swiss cross on the rudder.

Since 1937 Fl Kp 7 had a shark as the Kp emblem. In Summer 1940 the aggressive shark was replaced with a more cheerful trout. Avenches, Summer 1940.

The trout on its back. (See also the photograph on page 122)

Fl Kp 7 was often of assistance to surrounding farmers during fieldwork. During an inspection a higher officer scolded the 7ner as a band of farmers. Aircraft painter Hans Träutlein hit back. Avenches, Spring 1941.

Fl Kp 8 was the Flower *Kompanie*. Kpl Franz Gygax, a graphic designer, was the creator of this unusual motif. Avenches, Summer 1940

The "Narziss" (J-360) is one of the few exceptions of which the aircraft number is known.

The motifs were applied to both sides and were identical. Understandably, while taking the picture one focused on the flowers, but did not document any Flz numbers.

185

A gold comet served as the Kp emblem for Fl Kp 9. This retouched photograph shows the comet in the form from Summer 1940.

In the service period of Spring 1941 under Hptm Rufer, on 25 April a new form of the comet was introduced. At the same time the Kp orchestra "Die Goldkometen" was established.

J-350 is pulled from cover. Avenches, Spring 1941.

The Kp emblem of Fl Kp 15 in an earlier form. The photograph emerged in Spring 1940 when Kp emblems on aircraft were still rare.

The eagle as the symbol of Fl Kp 15 was introduced by the then Kp Kdt Hptm Herzig in 1937. It was applied in this form in Summer 1940. With the introduction of the DH 100 Vampire in 1955 the famous "Papierflieger" emerged.

The Kp emblem of Fl Kp 15 in Summer 1940.

187

The shark painting of Fl Kp 21 in Summer 1940. Notice the arrangement of the teeth.
The shark nose was seen for the last time in Spring 1952 on the Mustang of Fl St 21.

Fl Kp 21 in Emmen, Summer 1940. The round national emblem was still applied to the underside of the wings on J-377. Behind this are J-356 and J-313.

In a later period of service the shark was displayed with more teeth. Date of the photograph possibly Winter 1943/44. In the foreground is J-345.

At the end of a service period the Kp emblems had to be removed again. Notice the oversprayed emblem of Fl Kp 15, Olten, Summer 1940.

In later years of active duty the painting of Kp emblems was less common. Often the Kp number was painted with a washable color. Emergency landing of Oblt Dannecker as a result of an engine malfunction. Forel, May 11, 1944.

Below: J-353 of Fl Kp 9, 1942.

What was published next...!

A picture of the Me 109 V9/D-IPLU from the Dübendorf aviation meeting was an inspiration for this "Swiss Me 109."

"Jasta 15," a photograph montage from the year 1939/40.

Also, this well known German propaganda photograph received sharp criticism for the Messerschmitt 109 of the Swiss *Flugwaffe*.

This symbol of Fl St 7, with the trout "Jaqueline" mentioned in many publications, was created in 1952, and has nothing to do with the epoch of the Me 109.

190

Illustration from the "Livre d'Or" of the *Offizierskasino* Payerne

Notice the signatures of General Guisan, Div Bandi, BR Wetter, and BR Kobelt.

Illustration from the "Livre d'Or" of the *Offizierskasino* Payerne

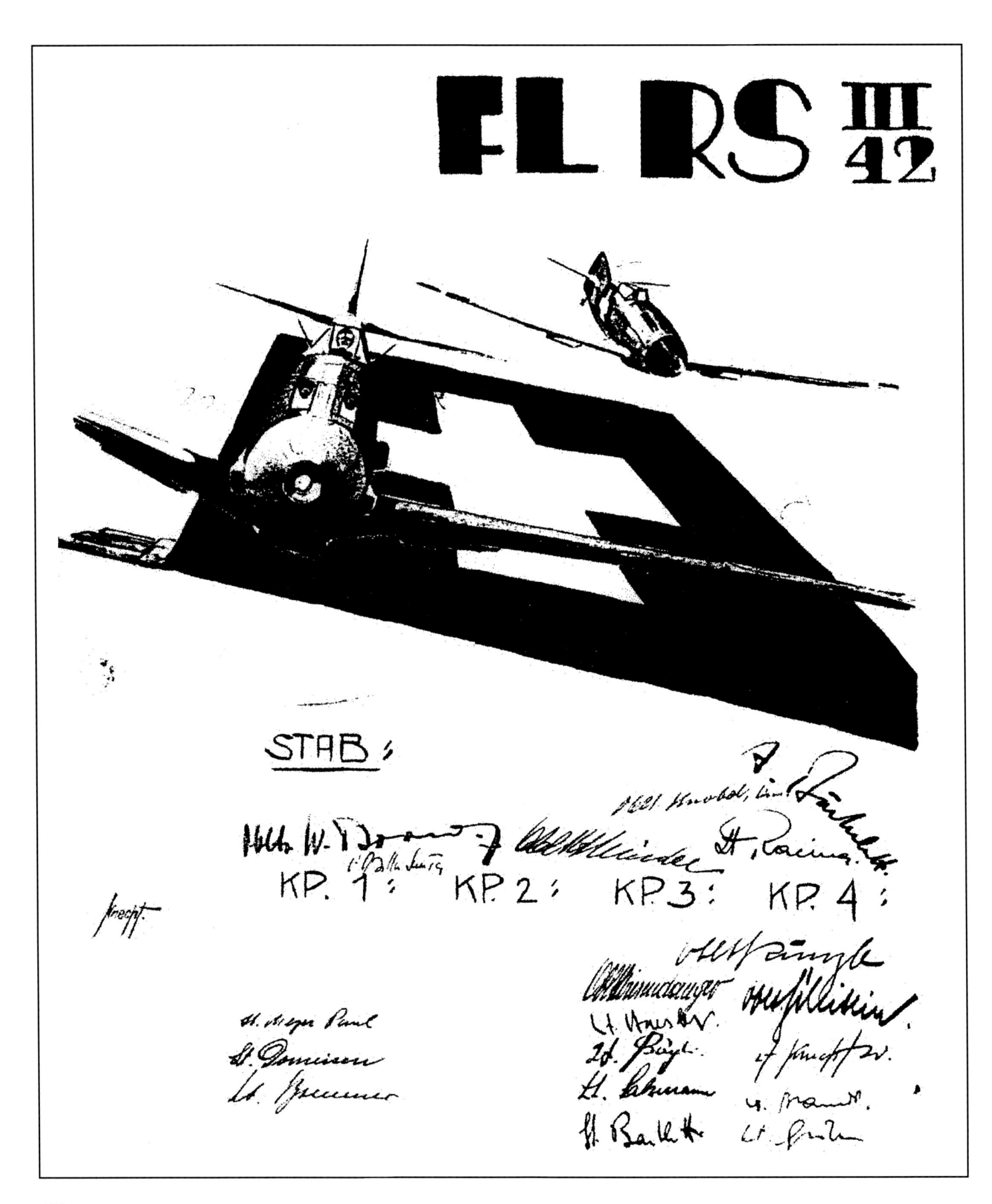

Chapter 17:

The Whereabouts of the Me 109 in Swizerland

The Pilatus P-2

During active duty, the Swiss *Flugwaffe* constantly required a greater number of training aircraft for advanced pilot training. At this time the Pilatus factory in Stans endeavored for a manufacture license for the Ambrosini S.A.I.7 and the Arado Ar 96 B. However, because the licensing laws were not granted, Pilatus developed a corresponding training aircraft on their own initiative.

The Pilatus P-2.01 flew for the first time on April 27, 1945. The Swiss *Flugwaffe* placed twenty-six P-2.05s in operation as of 1947. As of 1949 the same number with MG 29 armored (P-2.06s) followed. From the liquidation of the Me 109E numerous components for the P-2 could be utilized:

- Main landing gear with wheels and breaks, as well as retractable cylinders
- Tail bumper (entire assembly group)
- Rudder pedal with brake mechanism
- Hydraulic pump
- Diverse instruments
- MG 29 (only P-2.06)
- These aircraft, powered by an Argus As 410 A2, remained in the inventory of the Swiss *Flugwaffe* until 1981. After the decommission numerous aircraft were given to museums, or auctioned to private interests. Pilatus P-2s are still seen today at aviation meetings, partially with a fanciless painting.

J-355 in the Dübendorf Fliegermuseum

The factory no 2422 is one of the oldest maintained remaining Me 109s of all production series worldwide.

The aircraft was acquired on November 7, 1939, by the KTA, and subsequently given over to the troops. At this time J-355 had 3.31 operating hours. The aircraft was in troop service in *Flieger-Staffel* 8 for the last time in September 1949. With the order from the EMD on December 28, 1949, to shut down the entire Me 109 *Flotte* J-355 was decommissioned. The aircraft reached a total of 323.01 operating hours.

Afterwards J-355 was exhibited at aviation meetings but was no longer flown. As of the end of June 1959 it stood for twenty years in the Lucerne Museum of Transport. Since Summer 1979 J-355 has been located in the Fliegermuseum Dübendorf.

In the time as an exhibition piece in Lucerne J-355 was painted in a correct color scheme, but provided with the wrong national emblem.

Today the aircraft correctly represents the identification markings from January to April 1940. The color scheme KW 3 (RLM 71) is, however, incorrect. Furthermore, the security markings on the propeller blades was not applied at this time. Technically J-355 corresponds to the condition during the decommissioning in 1949 (cockpit equipment, assembly gaps for bomb equipment, and rockets and radio installation, as well as further external details that are not apparent).

A Pilatus P-1.05. The landing gear, as opposed to the Me 109, was pulled inwards.

J-355 above Lake Neuchâtel. Fl Kp 9, Summer 1942.

The cowling over the engine and the MG 29 corresponded to those of the Swiss reproductions J-392 to J-399.

Compromises were made for the presentation of aircraft from each era of active service. The three exhibited aircraft from the time of active duty represented the eras Spring 1940 (Me 109E), April to September 1944 (C 3603), and September 1944 until the end of the war (D 3801).

Assemblies, Weapons, and Engines

After the decommissioning of the Me 109G, the question was presented of a further utilization of the MG 131 for the defense of airfields.

At this time the army procured over 150 *Jagdpanzer* 38 (t) Hetzer at the Skoda factories in Pilsen. In Switzerland these were called *Panzerjäger* G-13.

In 1947 the KTA tested if the MG 131 could be installed in the G-13.

By means of the available information these weapons were, however, no longer used.

Firearms of the Me 109 like the MG 29, FF-K, MG 131, and MG 151/20 are exhibited today in the Fliegermuseum Dübendorf.

A DB 601 (No. 10768) and a DB 605 (not of a Me 109) are found in the engine hall of Dübendorf.

Before J-355 was permanently exhibited in the museum of transport, it was often seen at weapons exhibitions and aviation meetings. However, it was never shown in flight.

Me 109 pilots of Fl Kp 7 in front of J-355 in the aviation museum, Dübendorf, Spring 1993. From the left: Horst Siegfried, Robert Heiniger, Fritz Wiesendanger, and Hans Rudolf Hofer (Tech Of).

As was already mentioned, numerous Me 109s were used as target objects on the flyer shooting range by Forel, on Lake Neuchâtel. The remains were afterwards used as scrap metal.

The right wing of J-395 was able to be rescued in 1982; today it is privately owned.

In 1992 the left wing of a Me 109G was rescued by the author. By means of the manufacture number it was finally added to J-704.

The wing, however, was not applied to a Me 109G procured from Germany in 1944. The presumption is obvious that it originally came from one of the aircraft of Bern-Beundenfeld that made an emergency landing.

The Remnants of J-310

The debris of J-310, which crashed on June 4, 1940, by Boécourt was rescued privately in 1989. Astoundingly, one found the fuselage MG together with munitions. The airframe was completely destroyed, but sections of the engine, weapons, and equipment survived. The remains of this aircraft are today under private ownership.

The wing of J-395 was recovered in Forel in 1982. Hansruedi Dubler (middle) was the initiator for this undertaking.

In Summer 1992 the wing of J-704 was recovered. It exhibited some bullet holes and heavy corrosion. With the aid of sand blasting the corrosion could be treated. The wing is today protected with a wash primer. Photograph: Payerne, February 1993.

The remains of J-310. After 50 years in the earth, many parts remained in astonishingly good condition. In the center of the photograph the type plate of Arado Warnemünde with the Factory No. 2305.

Chapter 18:

Color Photographs

Two Me 109Es for the Swiss *Flugwaffe* at the factory airfield in Regensburg in October-November 1939. To the right in the photograph, Flugkapitän Trenkle.

Photograph below and the following page:
Aircraft of Fl Kp 9 in Avenches, Summer 1940. J-326 was given to the troops on June 27, 1939. On December 28, 1949, it was written off.

On October 26, 1939, J-350 arrived in the troop. During an interception maneuver after a bomb dropping in October 1946 the wings were deformed. J-350 subsequently had to make an emergency landing in Locarno. The aircraft was written off with 365 operating hours.

J-329 in Buochs. On June 28, 1939, the aircraft was given over to the *Fliegertruppe*. In June 1940 it suffered damages from German aircraft in air combat over the Jura. On August 14, 1948, J-329 was written off with 362 operating hours.

J-398 in the post-war period. The aircraft was given over to the troop on January 10, 1946. Notice the EW propeller.

Below: A Me 109E of Fl Kp 21 at the Buochs airfield, probably in Winter 1943/44.

These are presumably the last photographs of Fl St 8 in the year 1949. Notice the aircraft do not have armor plates in the cockpit. J-397, a Me reproduction, does not have an EW propeller.

J-303 Oblt Mooser, KTA
Dübendorf, December 1938

J-306
Dübendorf, Spring 1939

J-316
Dübendorf, June 1939

J-333, Fl Kp 21
Dübendorf, August 1939

J-323, Cp av 6
Thun, Spring 1940.

J-349, Fl Kp 21 Oblt Streiff
Dübendorf, May 1940.

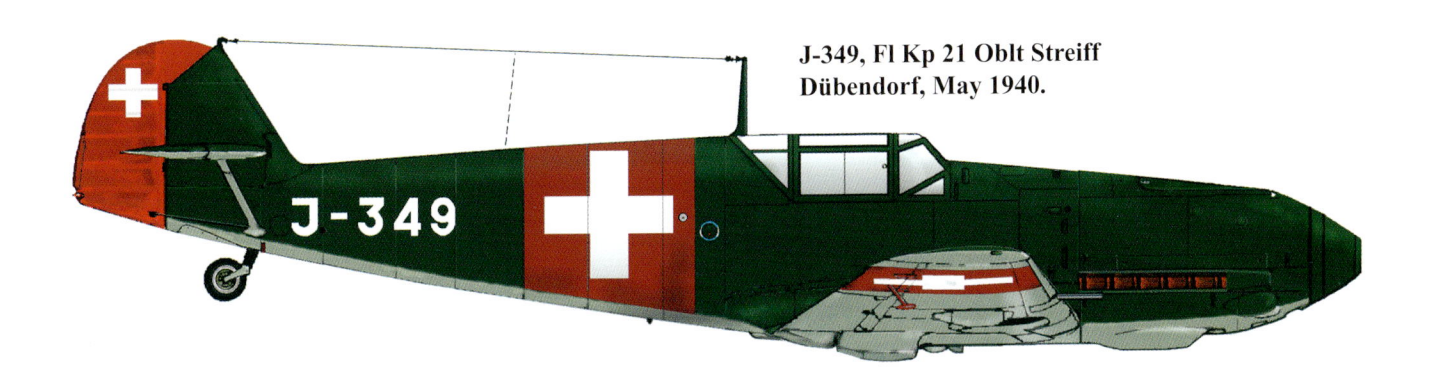

J-328, Fl Kp 15 Oblt Homberger
Olten, June 1940.

J-310, Fl Kp 15
Olten, June 1940.

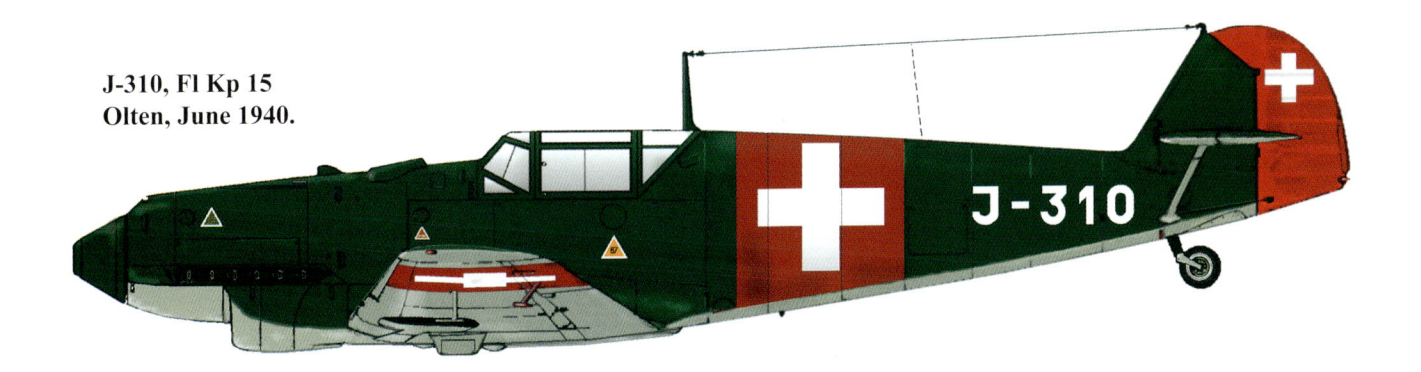

J-315, Cp av 6
Thun, Summer 1940.

J-330, Fl Kp 7
Avenches, July 1940.

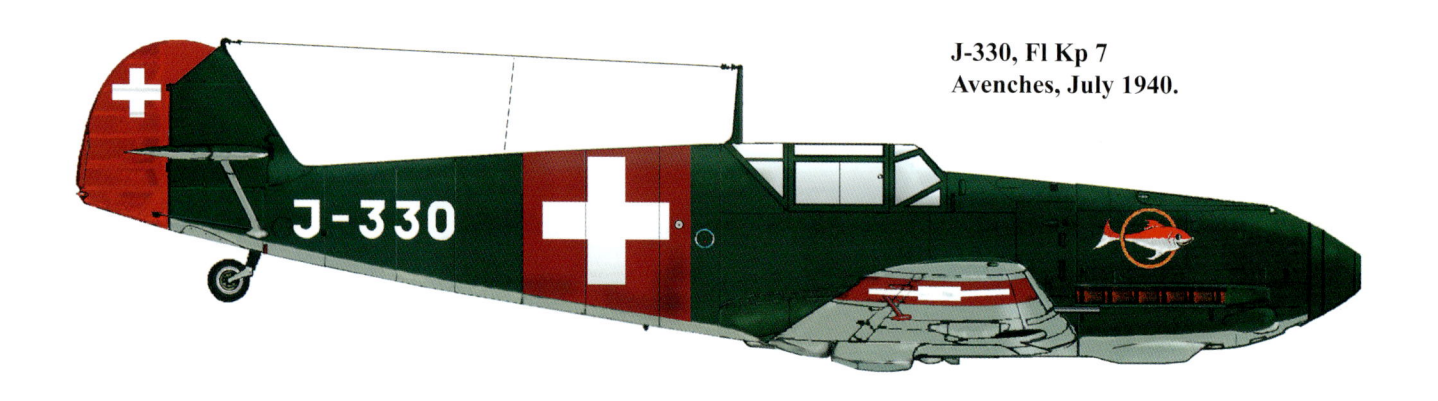

J-387, Fl Kp 15
Olten, Summer 1940.

J-353, Fl Kp 21
Emmen, Summer 1940.

202

J-360, Fl Kp 8
Avenches, Summer 1940.

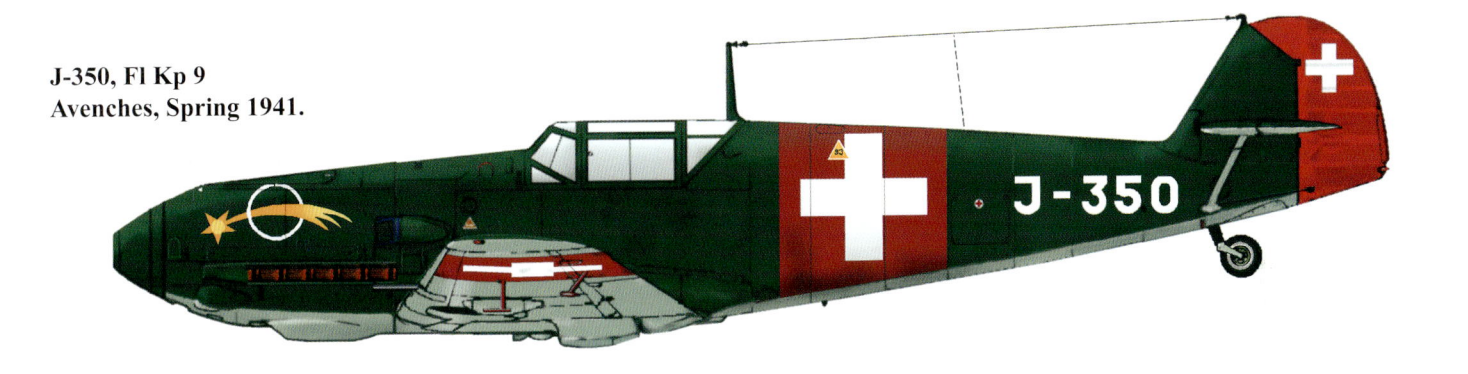

J-350, Fl Kp 9
Avenches, Spring 1941.

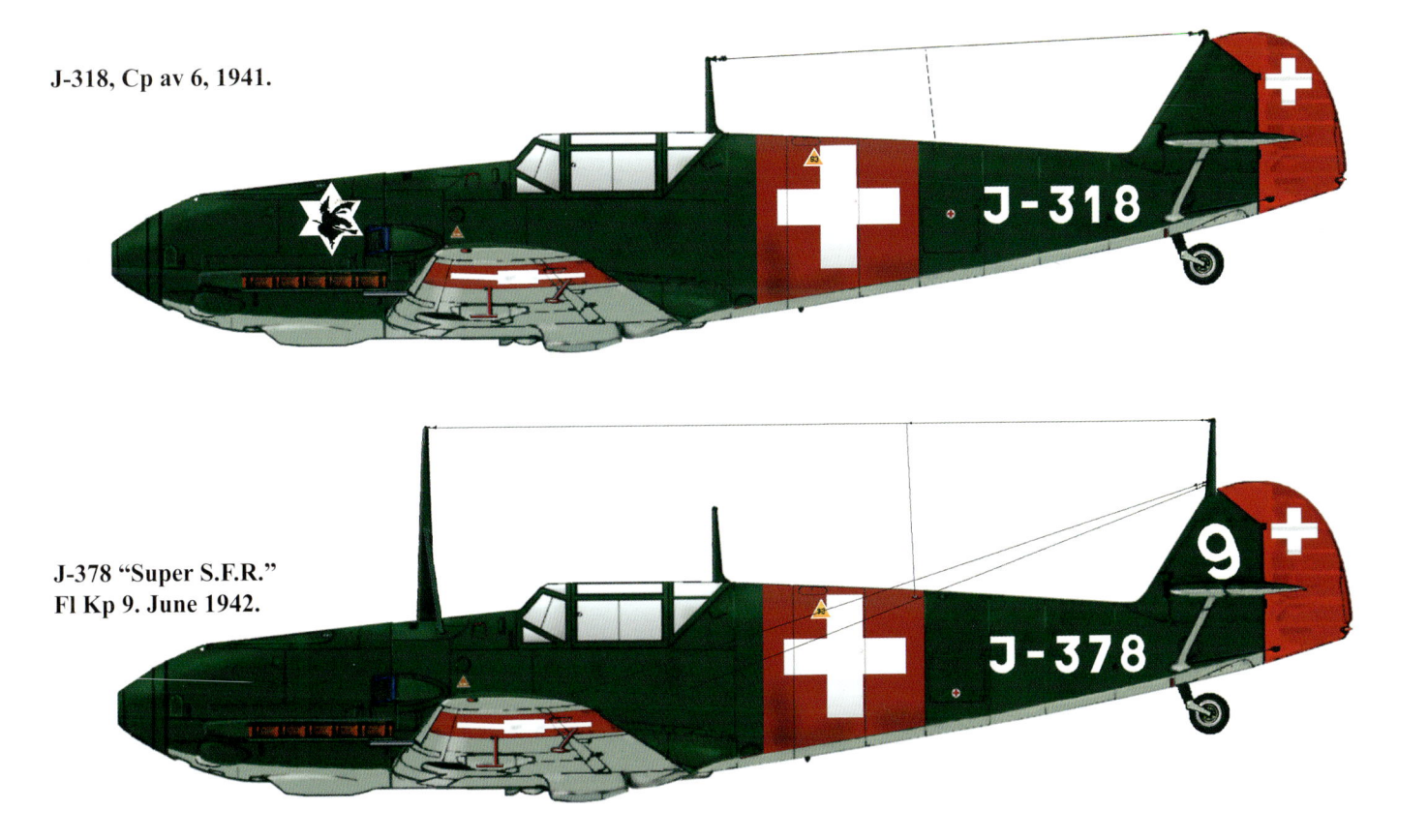

J-318, Cp av 6, 1941.

J-378 "Super S.F.R."
Fl Kp 9. June 1942.

203

J-702, Fl Kp 7
Interlaken, Spring 1944.

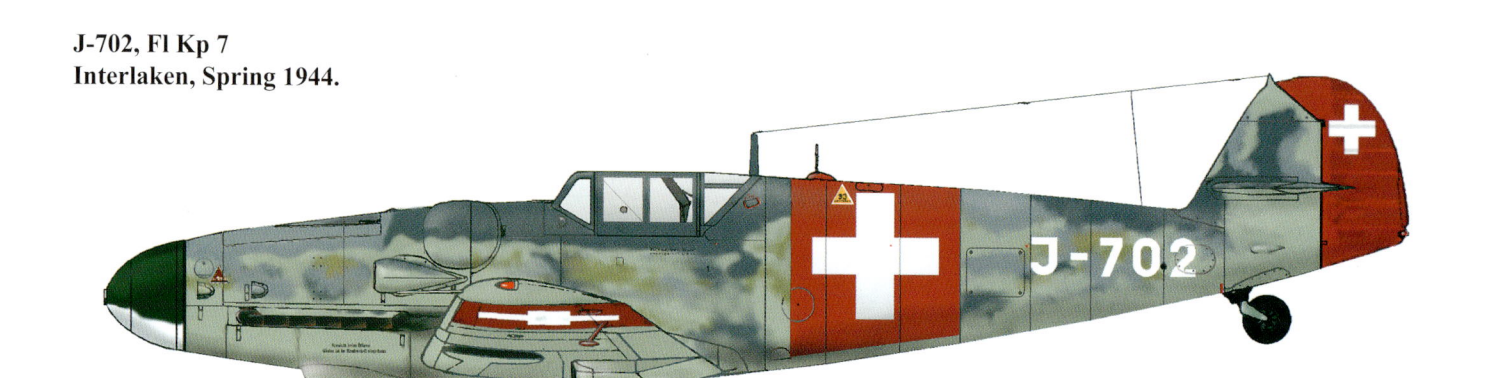

J-708, Fl Kp 7
Thun, Summer 1944.

Me 109 F Werk Nr. 7605
July 1943.

J-346, Fl Kp 8
Thun, Autumn, 1944.

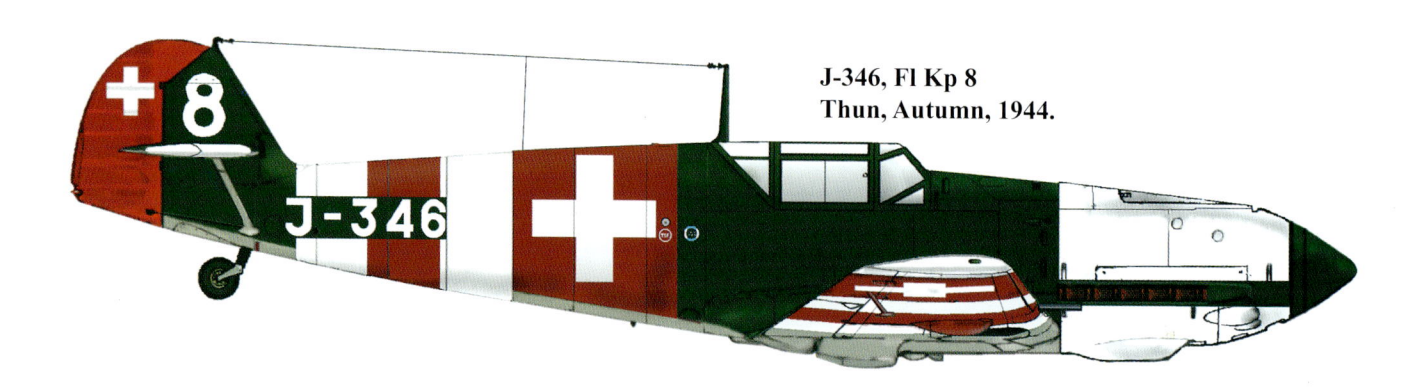

J-705, Fl St 7 Oblt Hofer.
Sion, November 1945

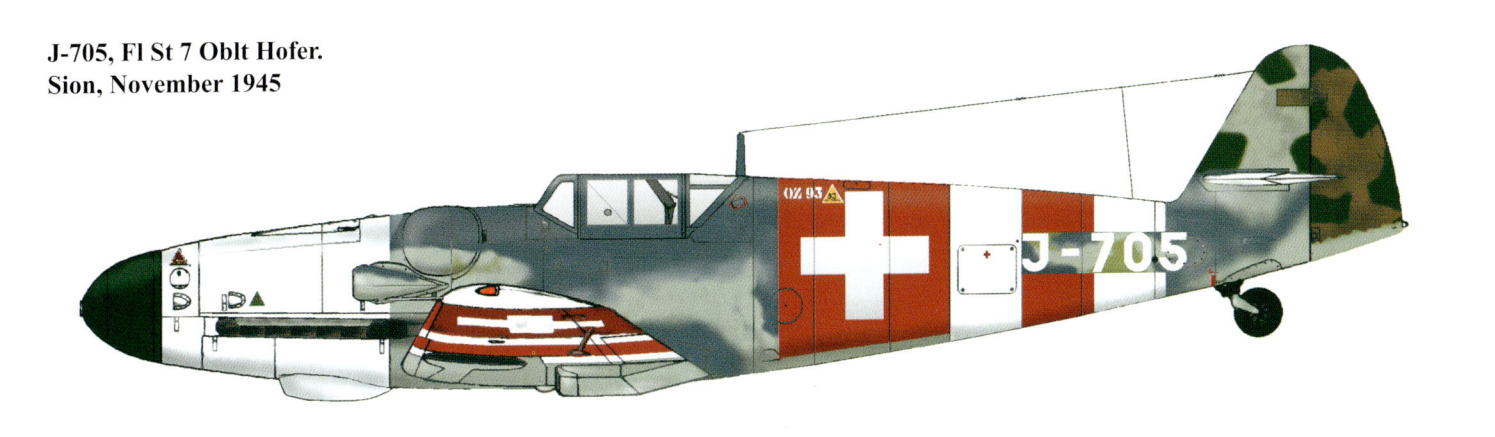

J-710, Fl St 7 Hptm Wiesendanger
Buochs, April 1946.

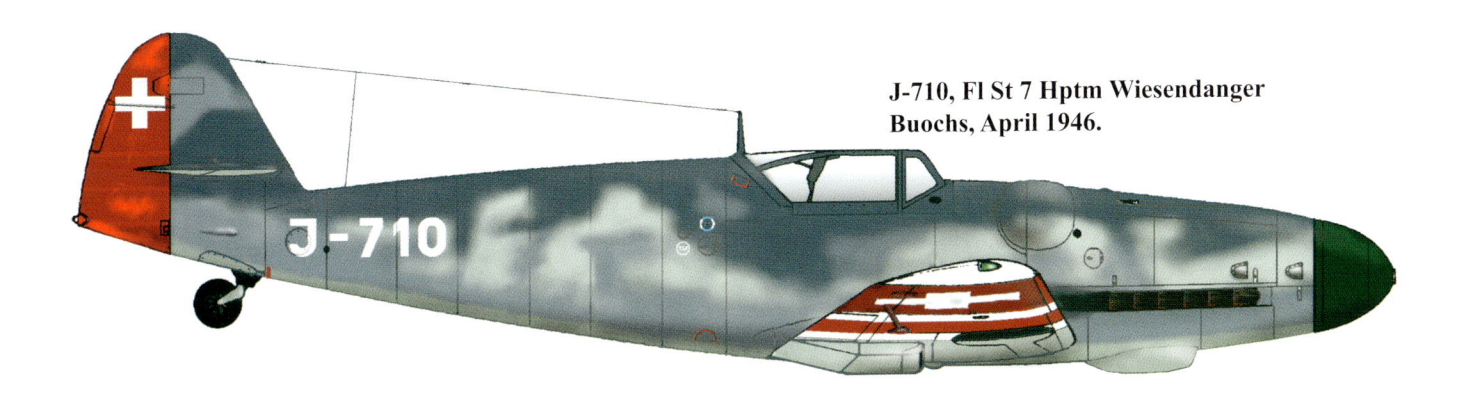

J-706
Buochs, June 1946.

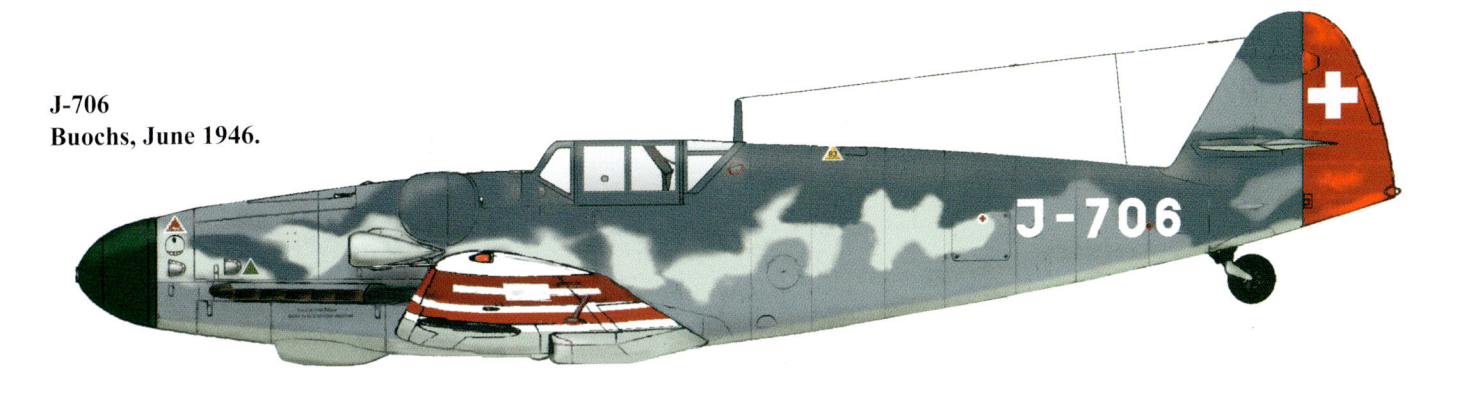

J-398
Fliegertruppe 1946-1949.

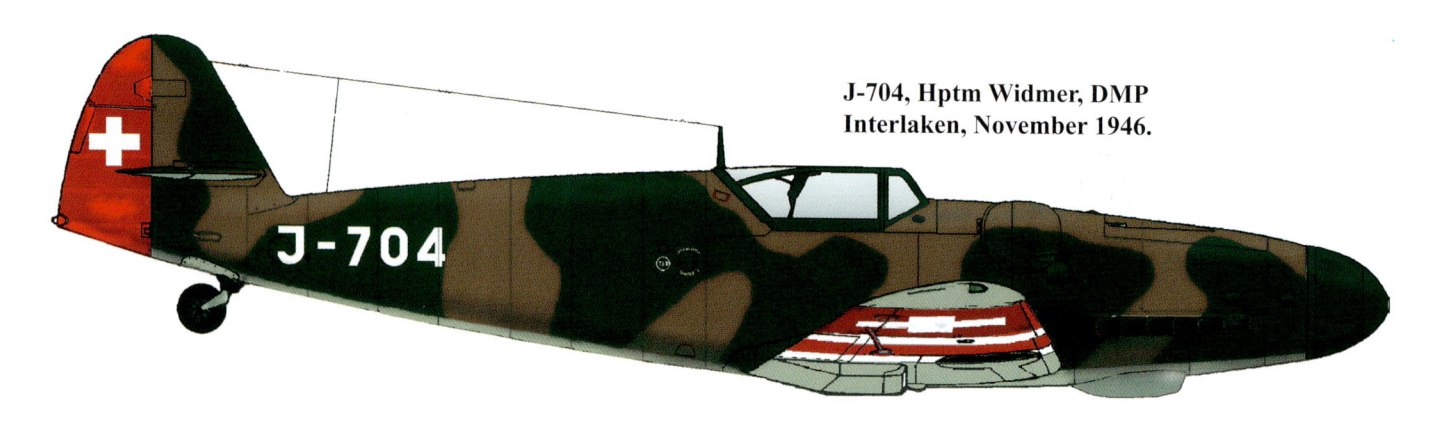

Camouflage Schema J-704
J-704, Fl St 7 Hptm Läderach
October 1945.

J-704, Hptm Widmer, DMP
Interlaken, November 1946.

Markings from 1939 to 1949

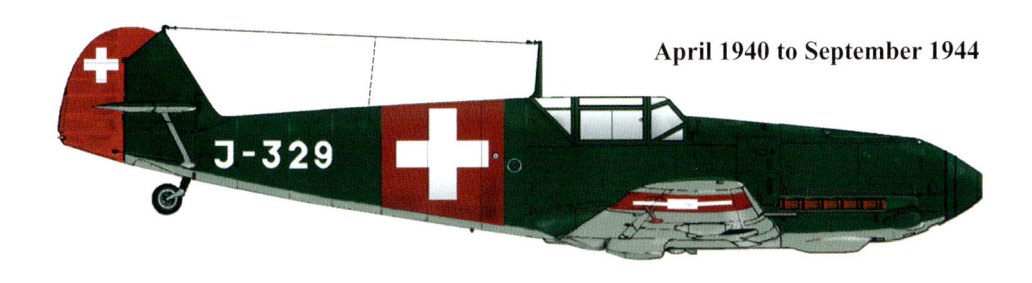

1939 to January 1940

January 1940 to April 1940

April 1940 to September 1944

September 1944 to September 1945

September 1945 to Withdrawal

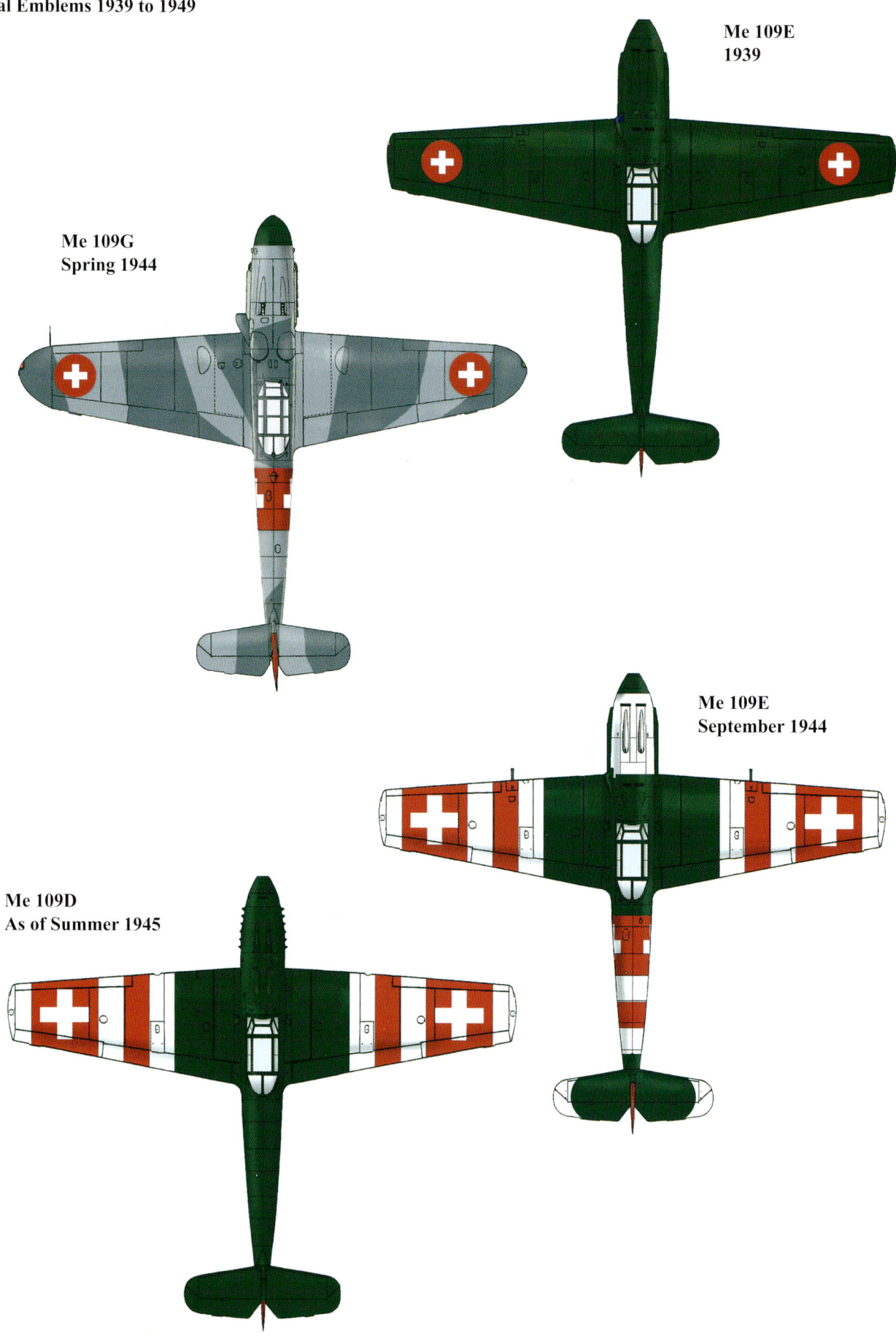

National Emblems 1939 to 1949

**Me 109E
1939**

**Me 109G
Spring 1944**

**Me 109E
September 1944**

**Me 109D
As of Summer 1945**

208

National Emblems on the Wings

Position above: 1938 to September 1944

Dimensions: Roundel 820 mm
Cross 470/145 mm

Position below: 1938 to January 1940

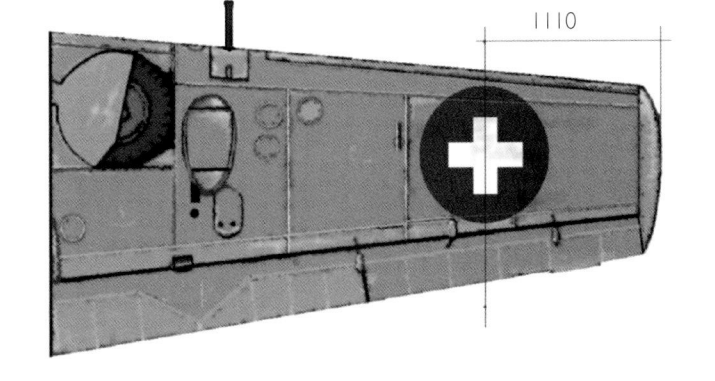

Position below: January 1940 to September 1944

Dimensions: Band 1300 mm
Cross 950/285 mm

Position of the Swiss cross according to instructions on January 13, 1940.
In practice the cross was positioned closer on the fore flap rear edge.

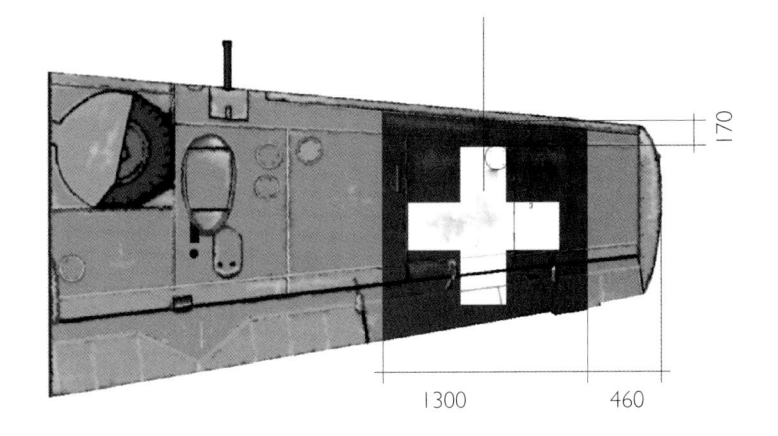

Position below and above:
September 1944 until decommissioning

Dimensions: Band 1300 mm
Cross 950/285 mm
Stripes 500 mm

The Swiss cross on the topside is flush with the fore flap rear edge.

After the end of the war identification stripes were removed from single aircraft.

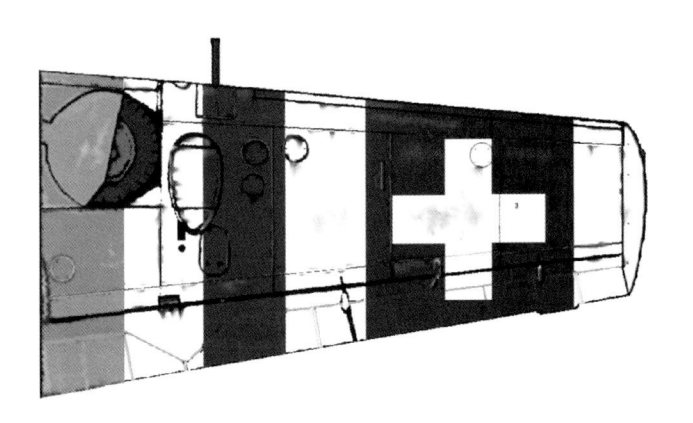

Markings and Aircraft Numbers

Me 109D

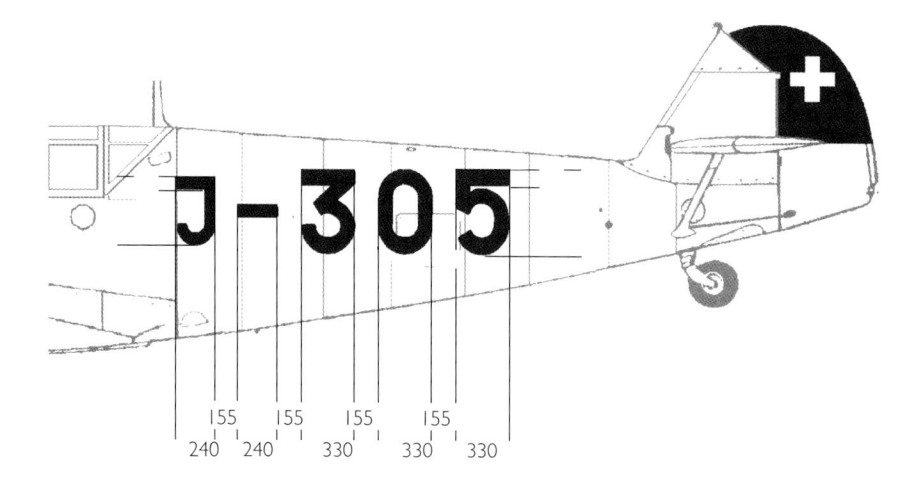

Me 109E

contour measurement to the fuselage middle

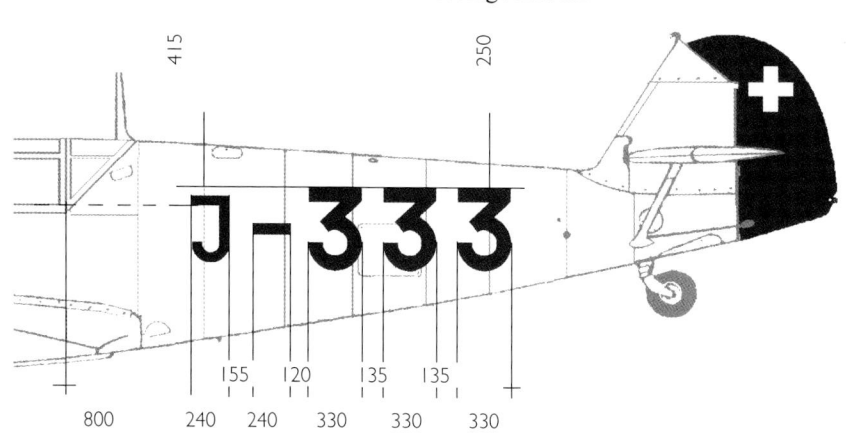

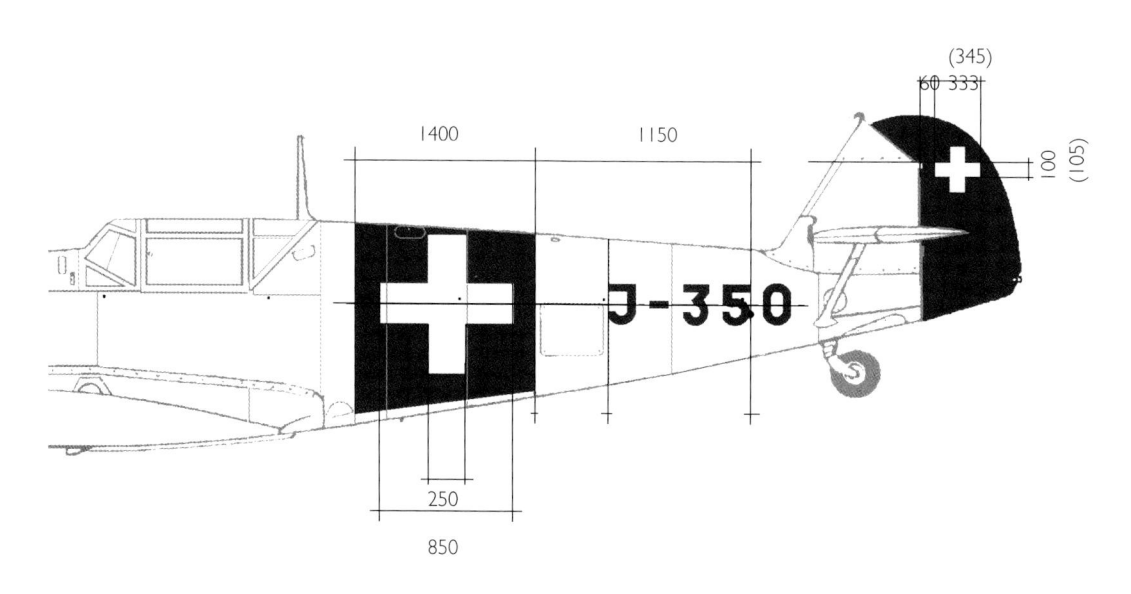

The horizontal centerline runs approximately on the top edge of the fuselage cover.

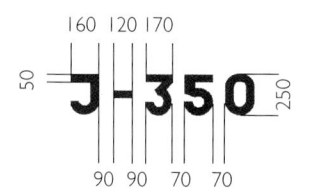

The Swiss cross on the rudder was often applied lower. The dimensions can vary.

210

The Me 109E Marking 1939/40

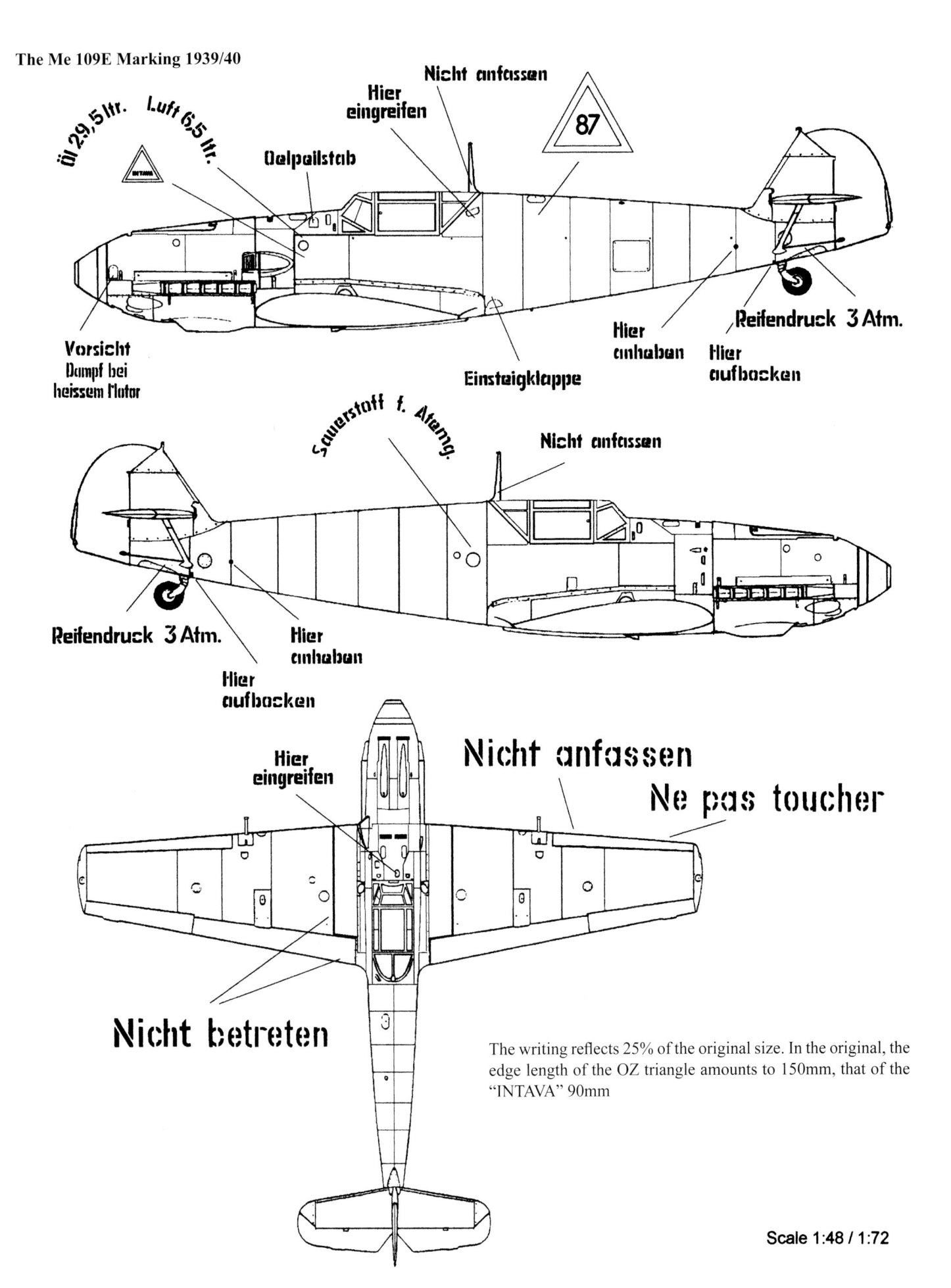

Öl 29,5 ltr. Luft 6,5 ltr.
Hier eingreifen
Nicht anfassen
Oelpeilstab
87
Vorsicht Dampf bei heissem Motor
Einsteigklappe
Hier anhaben
Reifendruck 3 Atm.
Hier aufbocken

Sauerstoff f. Atemg.
Nicht anfassen
Reifendruck 3 Atm.
Hier anhaben
Hier aufbocken

Hier eingreifen
Nicht anfassen
Ne pas toucher
Nicht betreten

The writing reflects 25% of the original size. In the original, the edge length of the OZ triangle amounts to 150mm, that of the "INTAVA" 90mm

Scale 1:48 / 1:72

211

Appendix 1:
Me 109D Weapon Installation MG 17

Because there are no records of the weapons assembly with the MG 29 available, examples from German sources with MG 17 are presented.

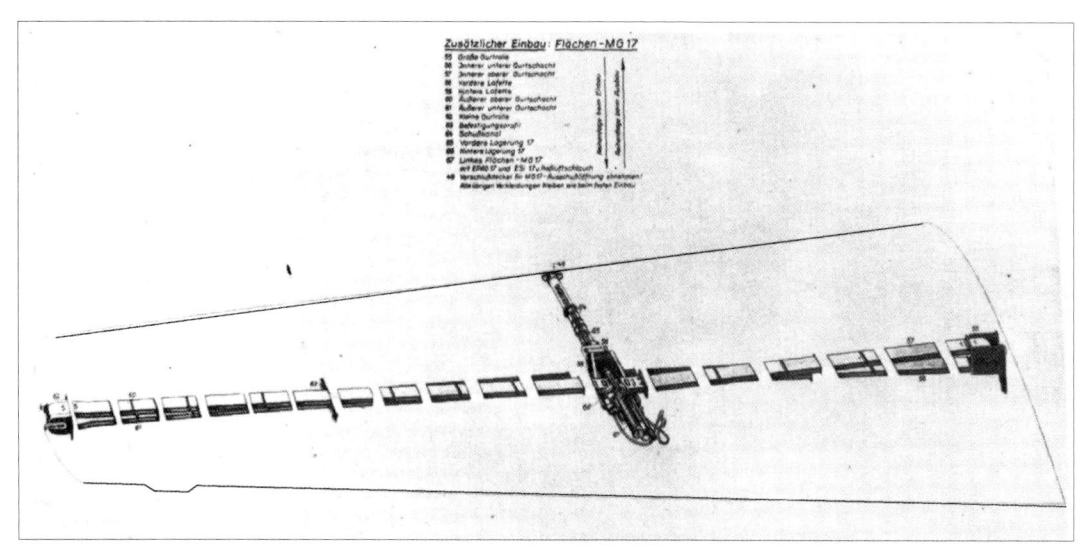

MG 17 with ammunitions belt over 500 cartridges
The MG 29 had at its disposal over 418 cartridges

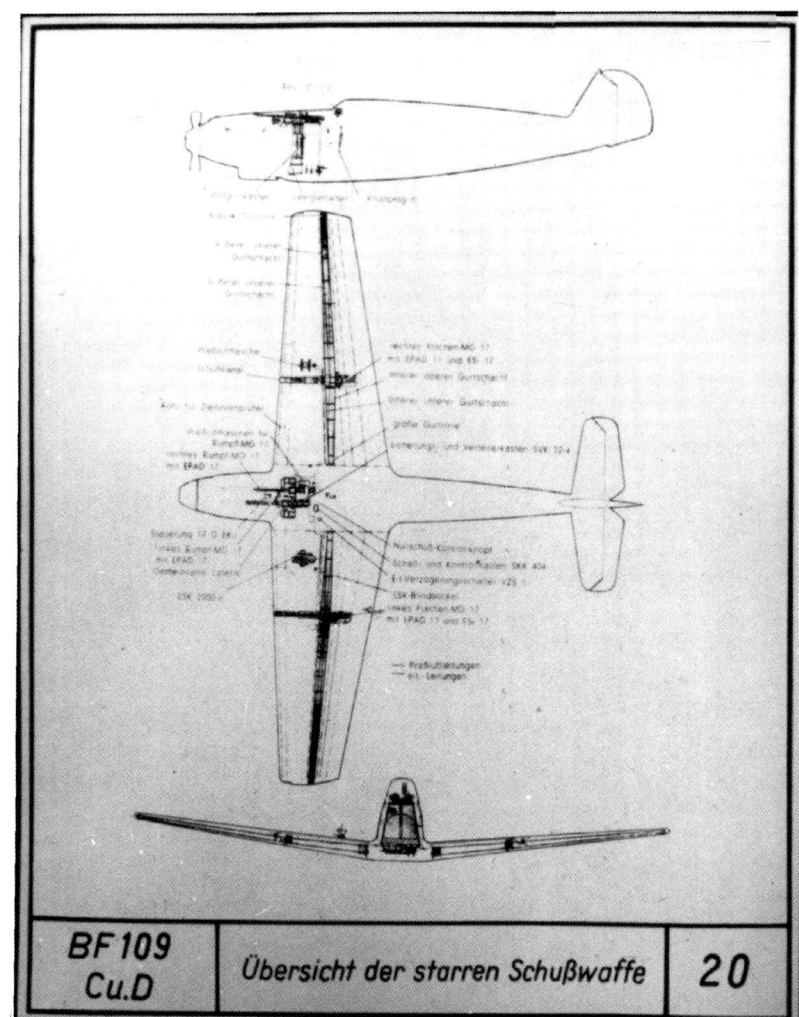

Weapons assembly with four MG 17s

The MG 17s above the engine were moved, the MG 29s installed parallel.

The 16 mm ESK 2000 film camera was not utilized in Switzerland.

Appendix 2:
Me 109E Cowlings of DB 601 and MG 29

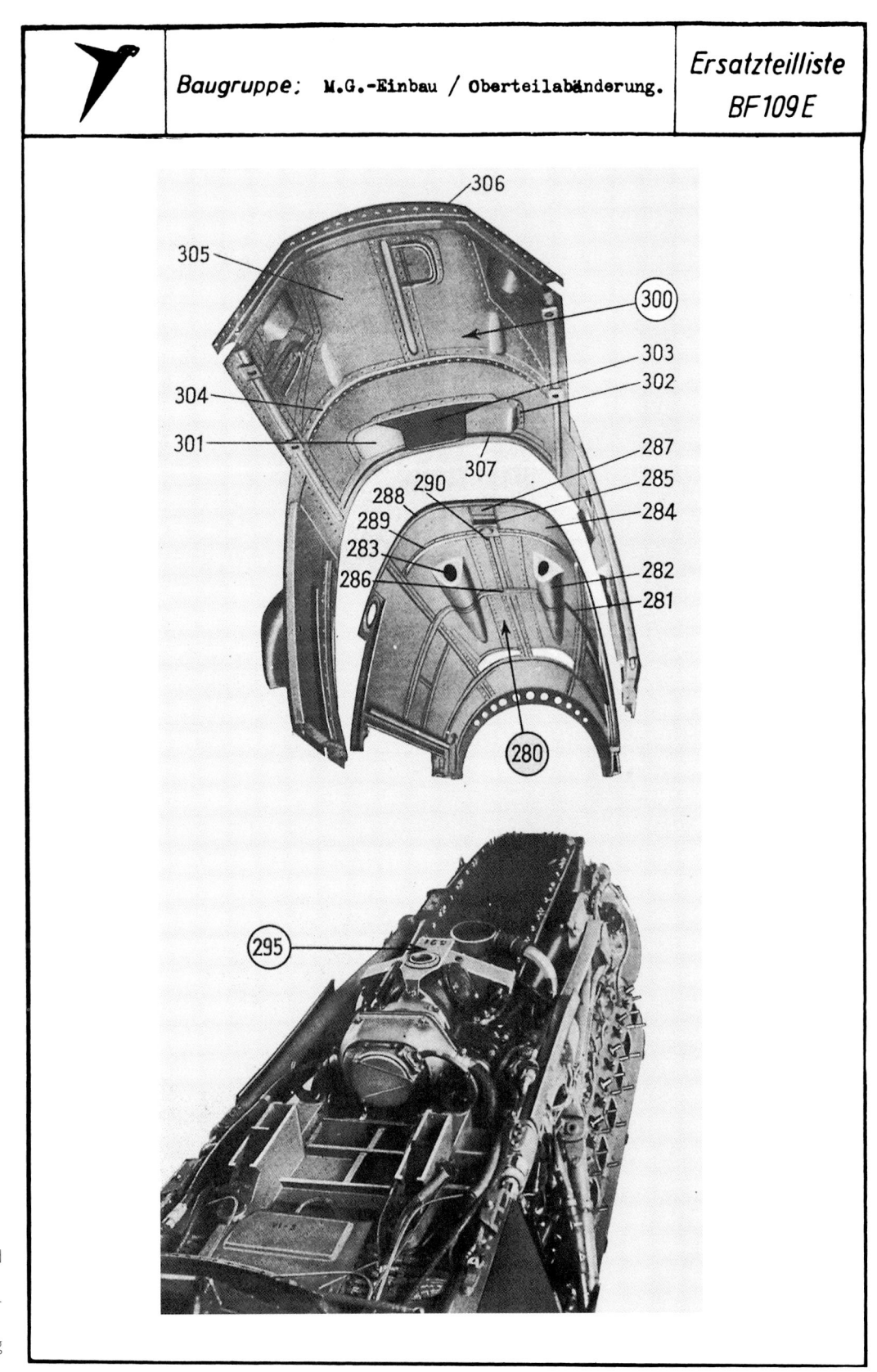

Baugruppe: M.G.-Einbau / Oberteilabänderung.

Ersatzteilliste BF 109 E

280 = anterior casing with changed bullet channels
300 = posterior casing with ventilation (301/302/303)
295 = plating for anterior casing (290)

Appendix 3:
Me 109E MG 29 - Weapons Installation

Baugruppe: Bewaffnungseinbau - Uebersicht M.G.	Ersatzteilliste BF 109 E

M.G. - Laffette	Fig. Nr.	1	Seite	401
Munitionskanäle	"	2	"	404
Gurtenkasten, Ableitkanäle, Hülsenkasten	"	3	"	406
Kabelzüge zum Pilotenraum	"	4*	"	411
Beschriftungen im "	"	5*	"	413
Verschalungsbleche über MG	"	6*	"	415
Synchronisierapparat	"	7	"	417

Appendix 4:
Me 109E MG 29 - Ammunition Box / Casing Box

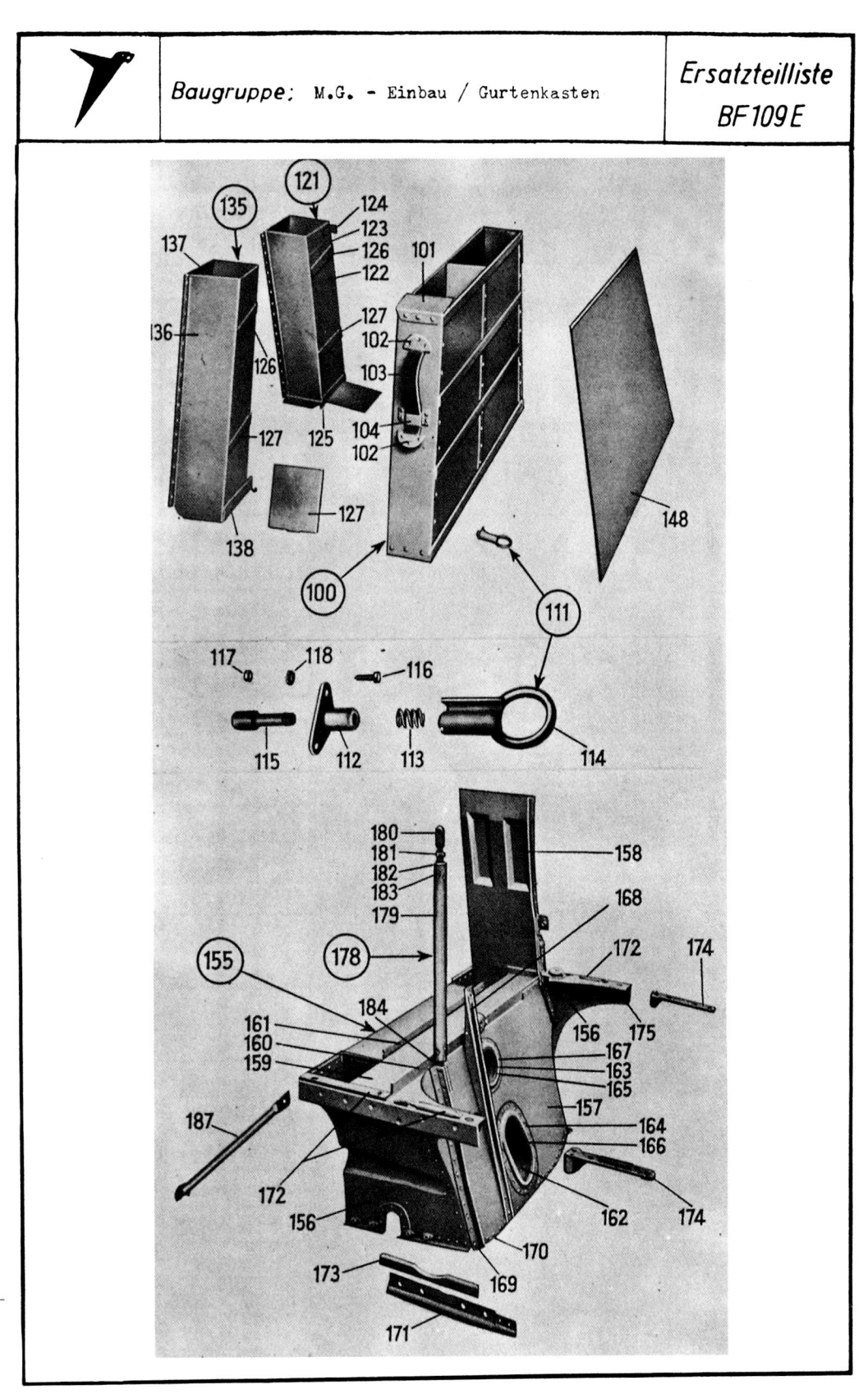

Baugruppe: M.G. - Einbau / Gurtenkasten

Ersatzteilliste BF 109 E

100 = ammunition box with two chambers to each 480 cartridges
121 = left shell discharge channel
135 = right shell discharge channel
155 = casing Box

Appendix 5:
Me 109E Control Unit for MG 29

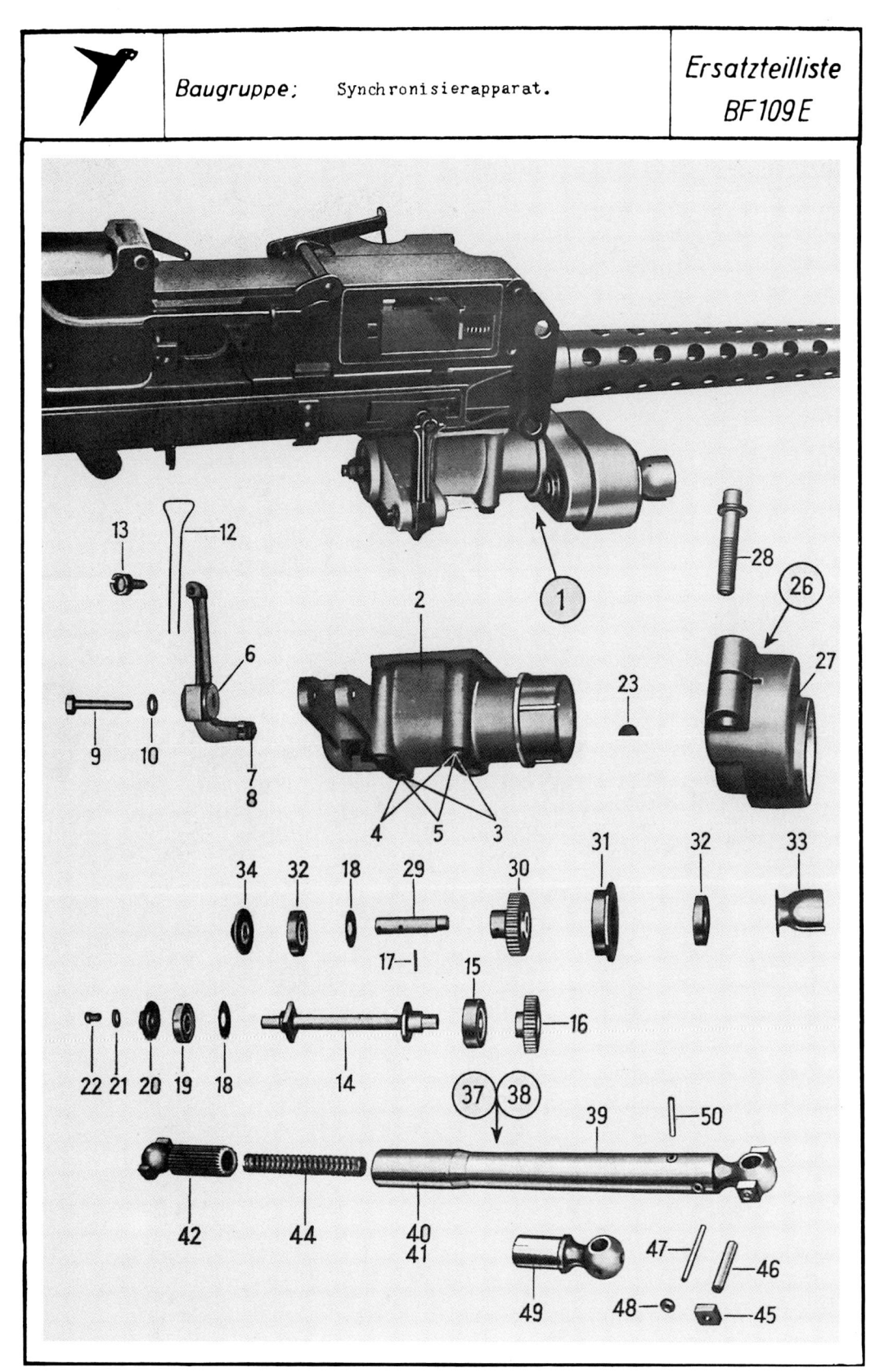

Baugruppe: Synchronisierapparat.

Ersatzteilliste BF109E

1 = synchronized apparatus with sideways shifted drive
2 = casing with mounting for the triggering lever (6)
37 + 38 = drive rod (synchronization shaft) with spherical head (42)

Appendix 6:
Me 109E Cable Pulley for MG 29

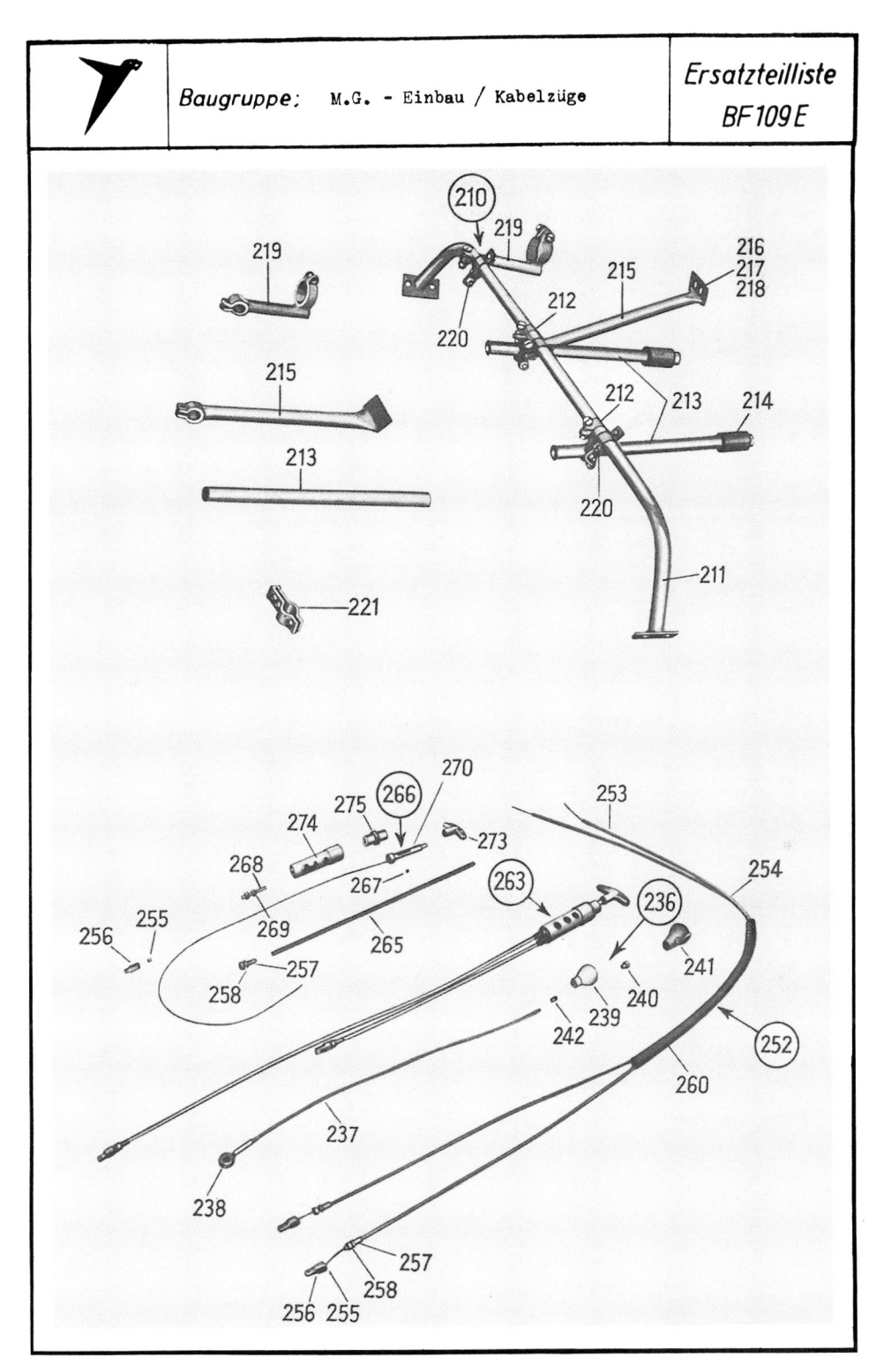

Baugruppe: M.G. - Einbau / Kabelzüge

Ersatzteilliste BF109E

210 = the frame with guide tubes and bracing
252 = MG trigger cable (led to the control stick)
236 = cable pulley for the leather knob (241) for loading movement (two cables ran through the tubes 213)

261 = cable pulley for the unloading movement (was fixed on pipe clamp 219)
The longer cable lead to the left weapon

217

Appendix 7:
Me 109E Assembly FF-K

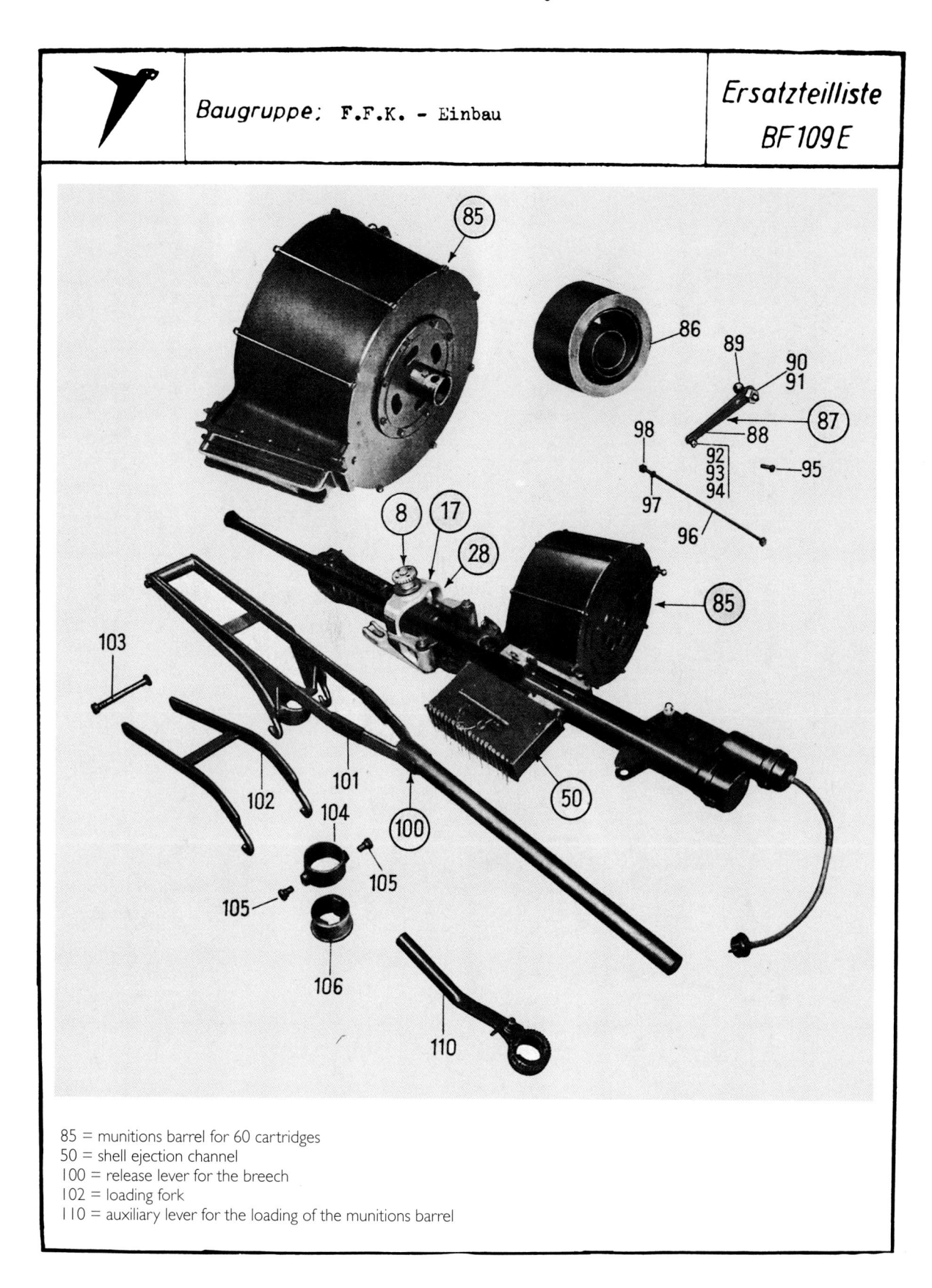

Baugruppe: F.F.K. - Einbau

Ersatzteilliste BF109E

85 = munitions barrel for 60 cartridges
50 = shell ejection channel
100 = release lever for the breech
102 = loading fork
110 = auxiliary lever for the loading of the munitions barrel

Appendix 8:
Me 109E Electrical Equipment for FF-K

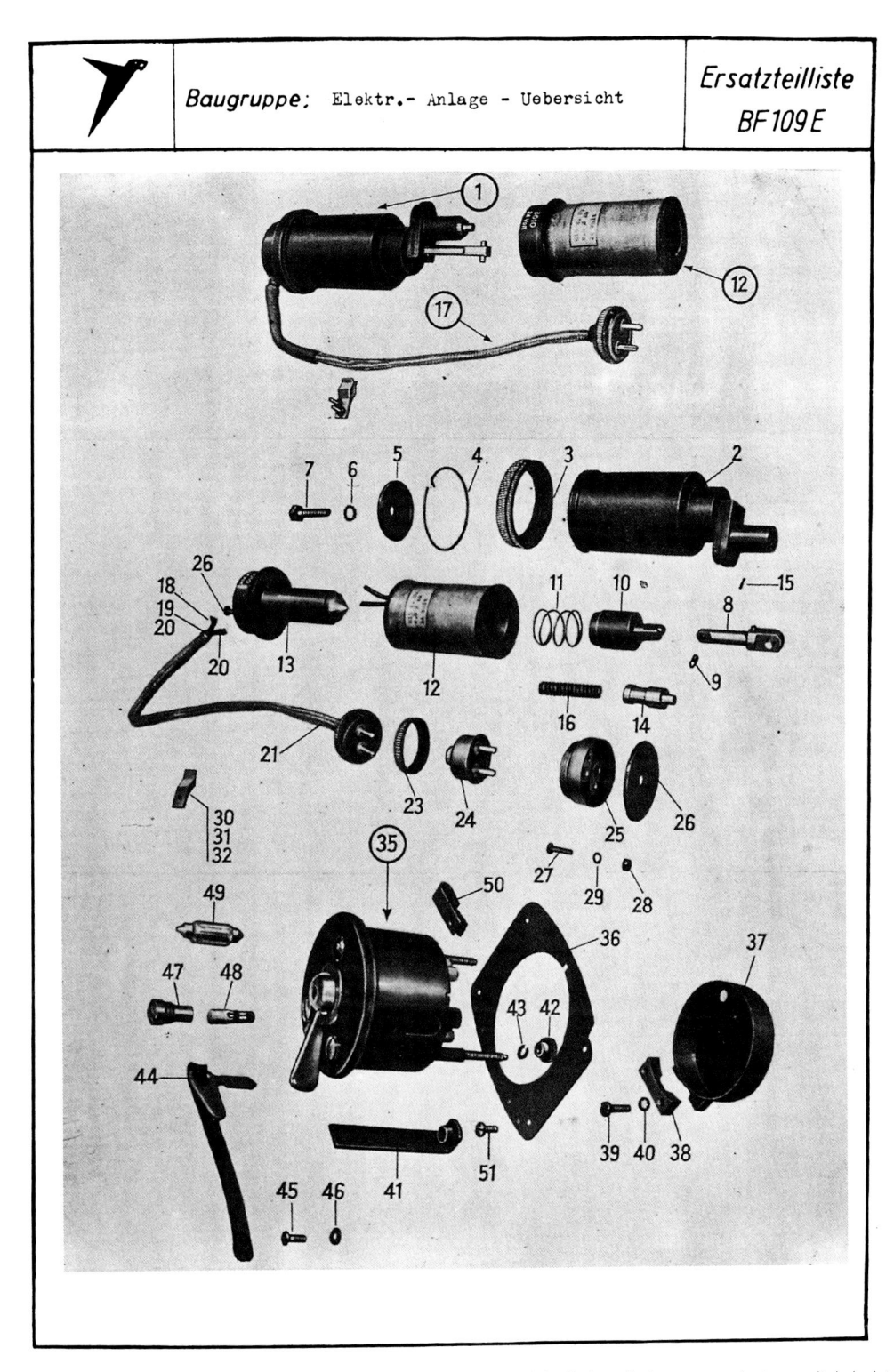

Baugruppe: Elektr.- Anlage - Uebersicht

Ersatzteilliste BF 109 E

1 = trigger magnet, attached to the cannon for the manipulation of the trigger mechanism
35 = cannon selector switch, original version with pilot light (49) and safety (47/48) for each cannon
44 = contact key with leather loop

The originally installed cannon selector switch led to problems of security (see page 46).

Definitive execution of the cannon selector switch, circa 1945.
Each cannon had a pilot light whose brightness could be regulated with a slide.

Appendix 9:
Me 109E Cable Pulleys for FF-K

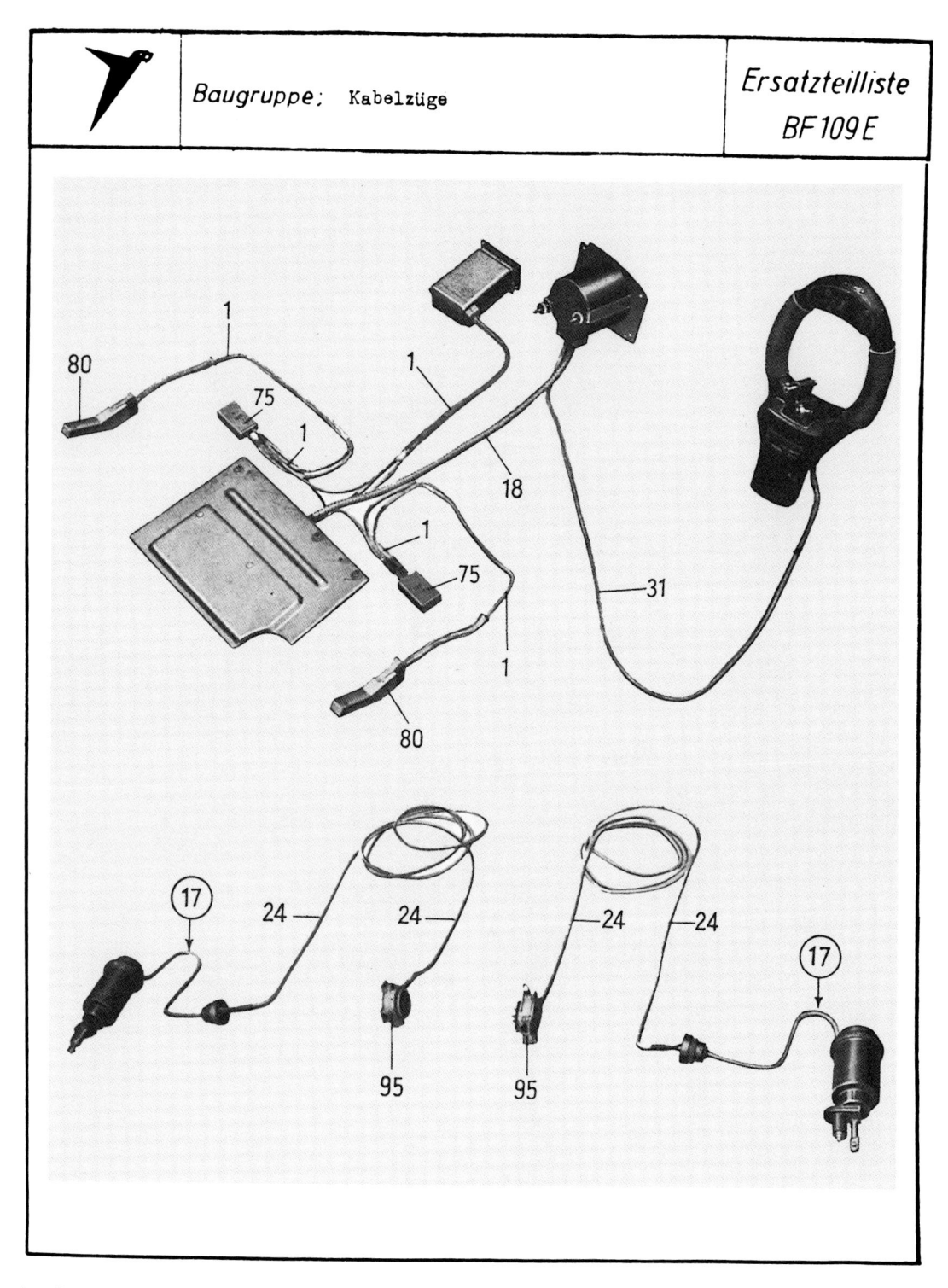

Baugruppe: Kabelzüge

Ersatzteilliste BF 109 E

1 = Cable to the shot counter, transmitter contacts (80) and condensers (75)
18 = Cable to the distribution box and cannon selector switch
31 = Cable to cannon selector switch and control stick KG 11
17 = Cable connection to trigger magnet

Appendix 10:
Me 109E Control Stick KG 11

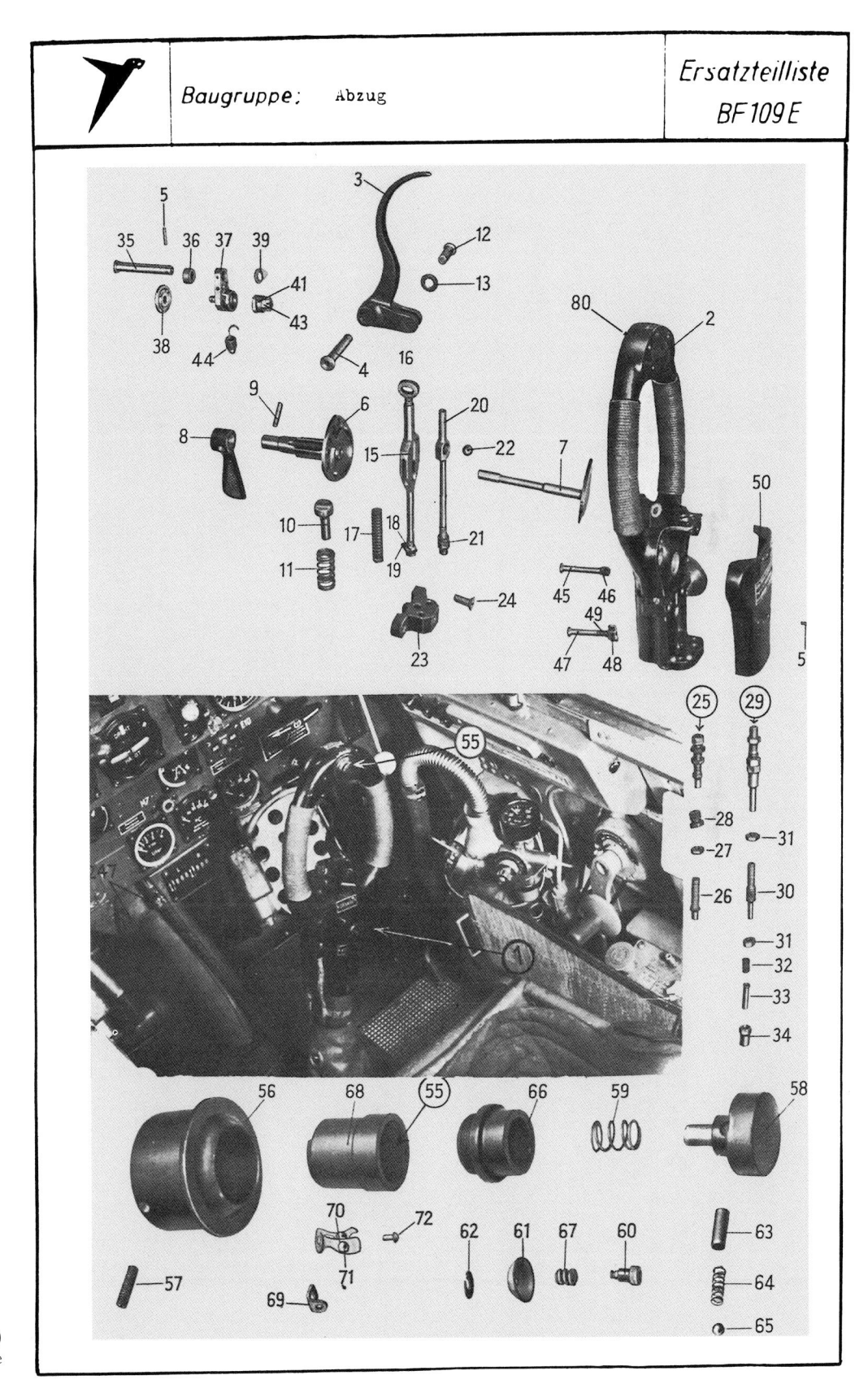

Baugruppe: Abzug

Ersatzteilliste
BF 109 E

2 = KG 11 grip
3 = MG 29 trigger arm (mechanical)
8 = MG 29 safety catch
55 (58) = FF-K trigger button (electrical)
80 = Push-button switch for bomb release (as of Summer 1944)

221

Appendix 11:
Me 109E Control Stick Contact Button

Armeeflugpark
Techn. Dienst
Nr.4111.13/T-5.Sc/m.

Interne Mitteilung Nr. 527

Selbstkontroll - Nr.

Me 109 E

Kontakttaste im Steuerknüppel Me 109 E.

Bei einigen Steuerknüppeln wurde die Feststellung gemacht, dass die Grundisolierplatten der Kabelführung zu den Kontakttasten der Schussauslösung und Bombenabwurfauslösung gebrochen waren.

Die nun losen Kontaktklemmen können einen Kurzschluss verursachen, was ein Auslösen der Schüsse oder Bomben zur Folge hätte.

Ich verfüge daher:

Bei der nächsten Kontrolle oder Revision, jedoch bis spätestens 31.12.44, sind die alten Kabel durch neue zu ersetzen, die bis zu den Kontakten der Auslöseknöpfe führen. Die dadurch hinfällig gewordenen Grundisolierplatten sind auszubauen.

Alte Ausführung!   Neue Ausführung!

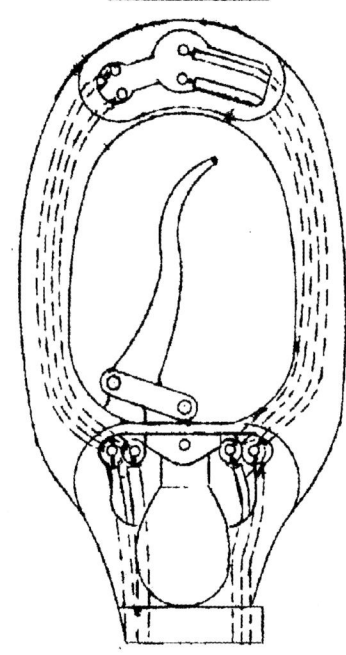

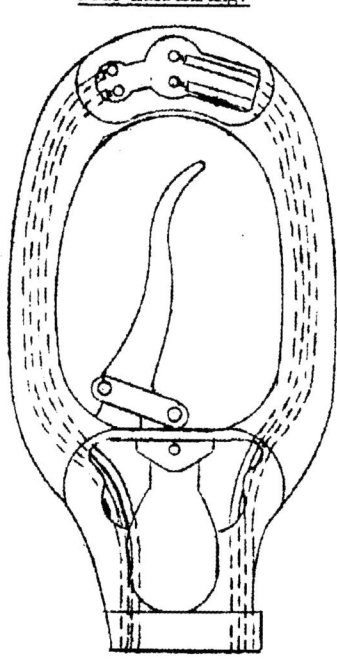

Ausgeführte Aenderungen sind mit Form. Nr. 77 dem Kdo. A.Fl.Pk., T-Dienst, zu melden.

Ausführung durch: Alle Gr. und Det.

Ausgabedatum: 16.11.44. Dringlichkeit: s.Text.

KDO. ARMEEFLUGPARK I. A. T-Chef:

Dringlichkeit: 1. Flz. darf ohne Ausführung obiger Weisung nicht geflogen werden.
2. Ausführung dringend, jedoch ohne Sperrung des Flz. für Flugdienst.
3. Ausführung bei nächster Revision.

Loose contact clamps could lead to an unwanted shot release or bomb dropping.
The electrical cables had to be replaced according to the IM No. 527.

Appendix 12:
Me 109E Aiming Device "Revi 3c"

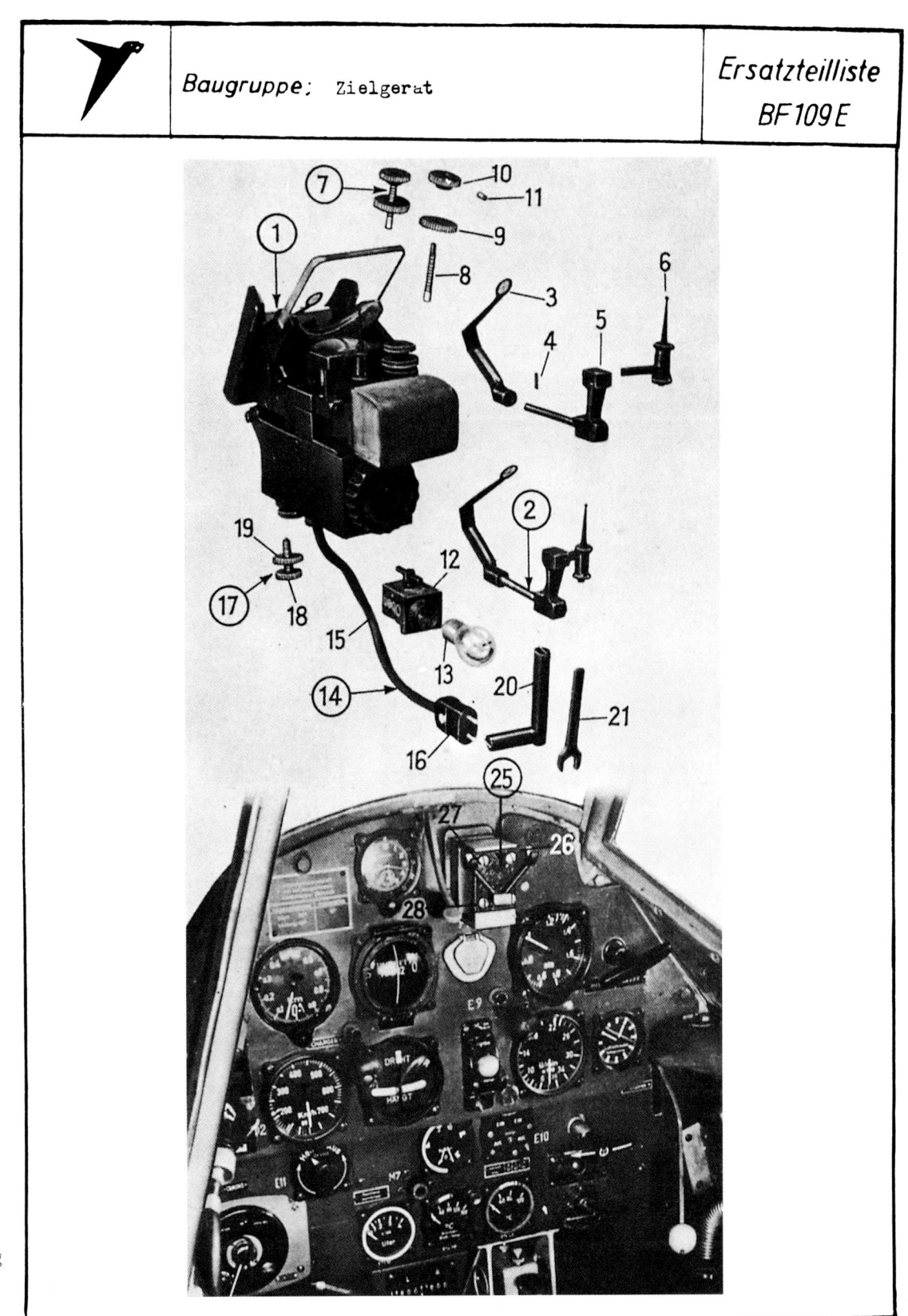

Baugruppe: Zielgerät

Ersatzteilliste
BF 109 E

1 = complete sight
2 = round sight with ring sight
20 = adjustable wrench
21 = wrench
25 = Revi support

Appendix 13:
Me 109E Signal Rocket Equipment

1 = Fold-out rocket box (between frame 6 and 7)
38 = Instrument panel in cockpit   - above pilot light
   - below release switch
   - master switch

Rocket boxes on museum aircraft J-355
The magazine could be loaded with a red, green, and white rocket.

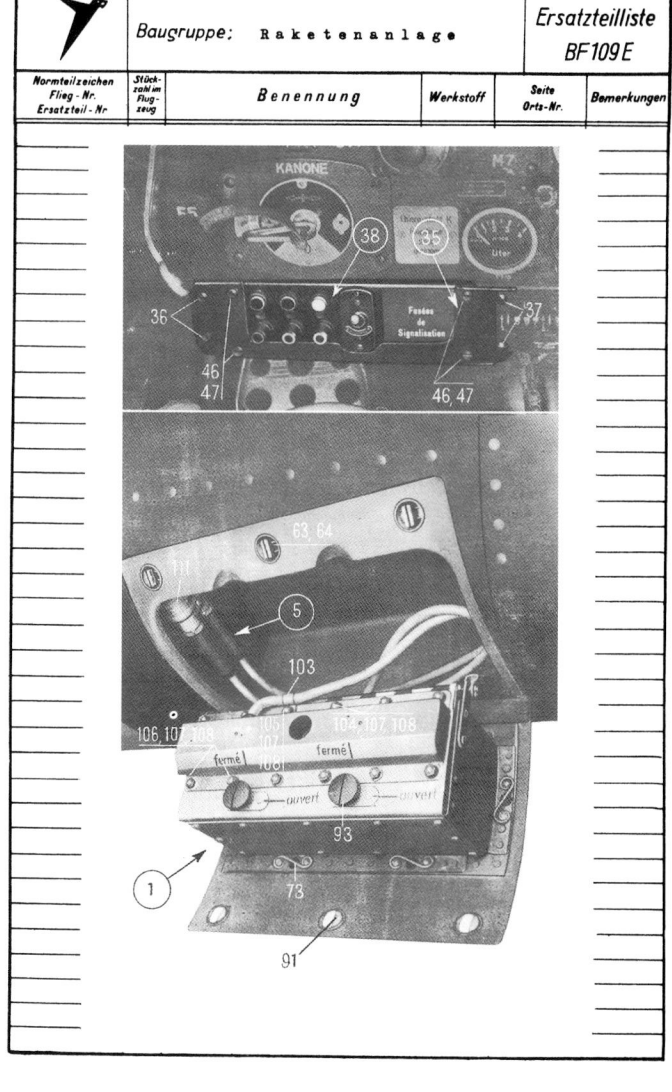

In equipment ready for operation the small cover (with the 6 rotary seals) was removed in order to clear the exit opening.

Appendix 14:
Me 109E Bomb Equipment

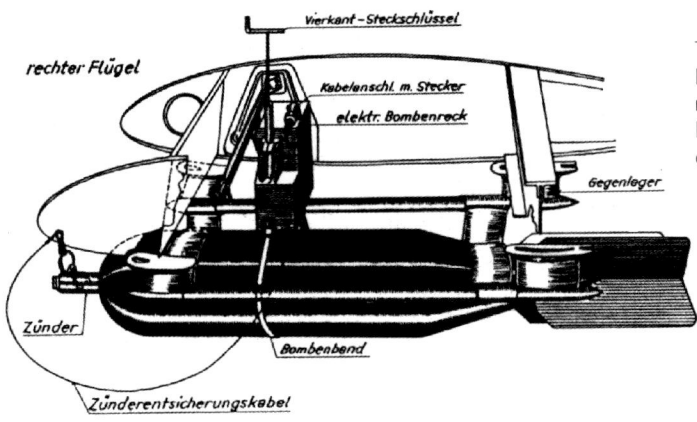

The release of the detonator occurred over a detonator release cable. During sharp release "explosif," the cable locked on the bar, and thereby released the bombs.
During a secured release "Inert" the bombs released together with the cable.

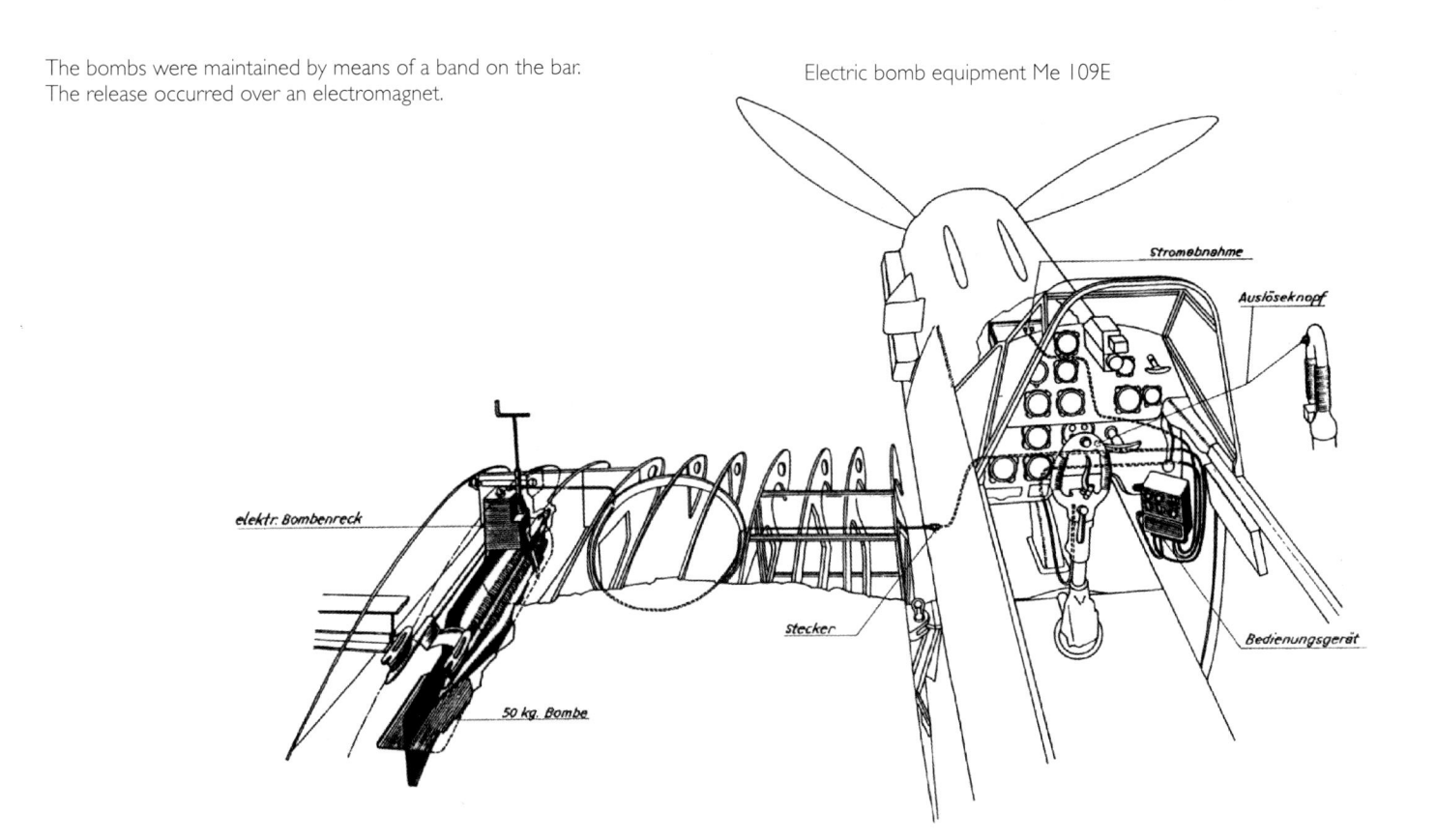

The bombs were maintained by means of a band on the bar. The release occurred over an electromagnet.

Electric bomb equipment Me 109E

Appendix 15:
MG 29 Adjustment Table for Synchronization

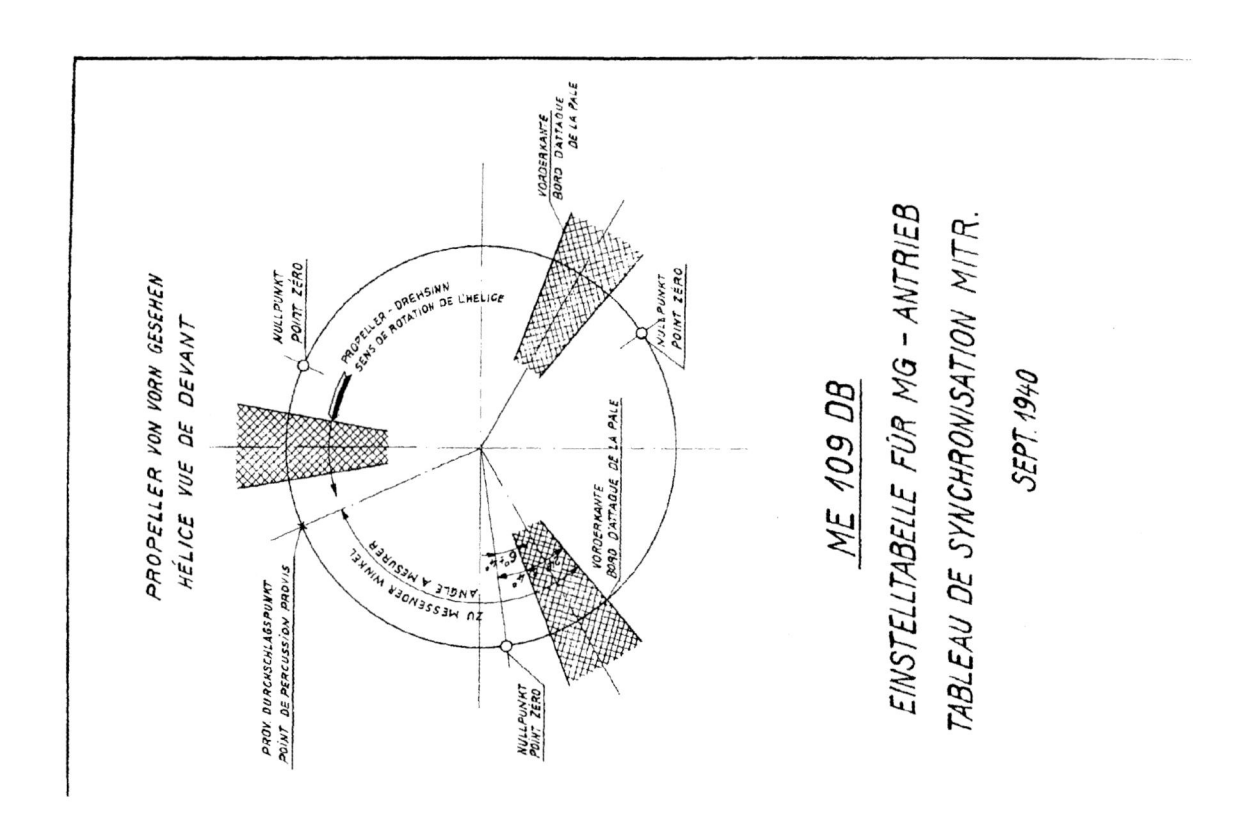

ME 109 DB
EINSTELLTABELLE FÜR MG - ANTRIEB
TABLEAU DE SYNCHRONISATION MITR.
SEPT. 1940

Verzögerung Retardement Prov. Einstellung Réglage prov.	Zahn Nr. - Dent Nr. Antriebs-Kopf Tête de commande	Stange Tige de commande	Verzögerung Retardement Prov. Einstellung Réglage prov.	Zahn Nr. - Dent Nr. Antriebs-Kopf Tête de commande	Stange Tige de commande	Verzögerung Retardement Prov. Einstellung Réglage prov.	Zahn Nr. - Dent Nr. Antriebs-Kopf Tête de commande	Stange Tige de commande
0	1	12	45	0	1	90	3	12
1	2	12	46	30	12	91	3	12
2	2	13	47	31	13	92	4	13
3	0	24	48	31	13	93	4	13
4	0	24	49	31	13	94	4	13
5	0	24	50	29	24	95	2	24
6	0	24	51	29	24	96	2	24
7	0	0	52	30	0	97	3	0
8	1	0	53	30	0	98	3	0
9	1	0	54	30	0	99	3	0
10	1	0	55	30	0	100	3	0
11	2	1	56	30	1	101	3	1
12	0	12	57	31	1	102	4	1
13	0	12	58	29	12	103	4	12
14	0	12	59	29	12	104	2	12
15	0	12	60	29	12	105	2	12
16	0	13	61	29	12	106	2	12
17	1	13	62	29	13	107	2	13
18	1	13	63	30	13	108	3	13
19	1	13	64	30	13	109	3	13
20	31	24	65	4	24	110	1	24
21	31	24	66	4	24	111	2	24
22	0	0	67	29	0	112	2	0
23	0	0	68	29	0	113	2	0
24	0	0	69	29	0	114	2	0
25	0	0	70	29	0	115	2	0
26	1	1	71	29	1	116	2	1
27	1	1	72	30	1	117	3	1
28	1	1	73	30	1	118	3	1
29	31	12	74	4	12	119	1	12
30	31	12	75	4	12			
31	31	12	76	4	12			
32	0	13	77	29	13			
33	0	13	78	29	13			
34	0	13	79	29	13			
35	0	13	80	3	24			
36	30	24	81	3	0			
37	31	0	82	4	0			
38	31	0	83	4	0			
39	31	0	84	4	0			
40	31	0	85	4	0			
41	0	1	86	4	1			
42	0	1	87	29	1			
43	0	1	88	29	1			
44	0	1	89	3	12			

Appendix 16:
Me 109E Comparison VDM – EQ V6 Propeller

Kdo.Fl.u.Flab.Trp.

Leistungsmessung mit Flz. Me 109 E

Vergleich zwischen VDM - und EW.V6 - Propeller

August 1942

CINA Höhe m	Geschwindigkeiten km/h Schraube VDM	Geschwindigkeiten km/h Schraube EW.V6	Steigzeit VDM	Steigzeit Schraube EW.V6
0	441	442	0' 0"	0' 0"
1000	460	461	1' 4"	1' 12"
2000	478	481	2' 8"	2' 14"
3000	497	501	3' 22"	3' 26"
4000	515	521	4' 48"	4' 40"
5000	532	291	6' 16"	6' 2"
6000	542	555	8' 6"	7' 34"
7000	540	559	10' 24"	9' 24"
8000	526	547	13' 48"	12' 6"
9000	500	526	19' 00"	16' 40"
10000	465	498	29' 20"	24' 40"

Rollänge beim Start in H = 600 m   mit VDM - Schraube   L = 287 m
                                    mit EW.V6 - Schraube L = 218 m

A/5 148x210

Appendix 17:
He 111 1G+HT, Kemleten, May 16, 1940

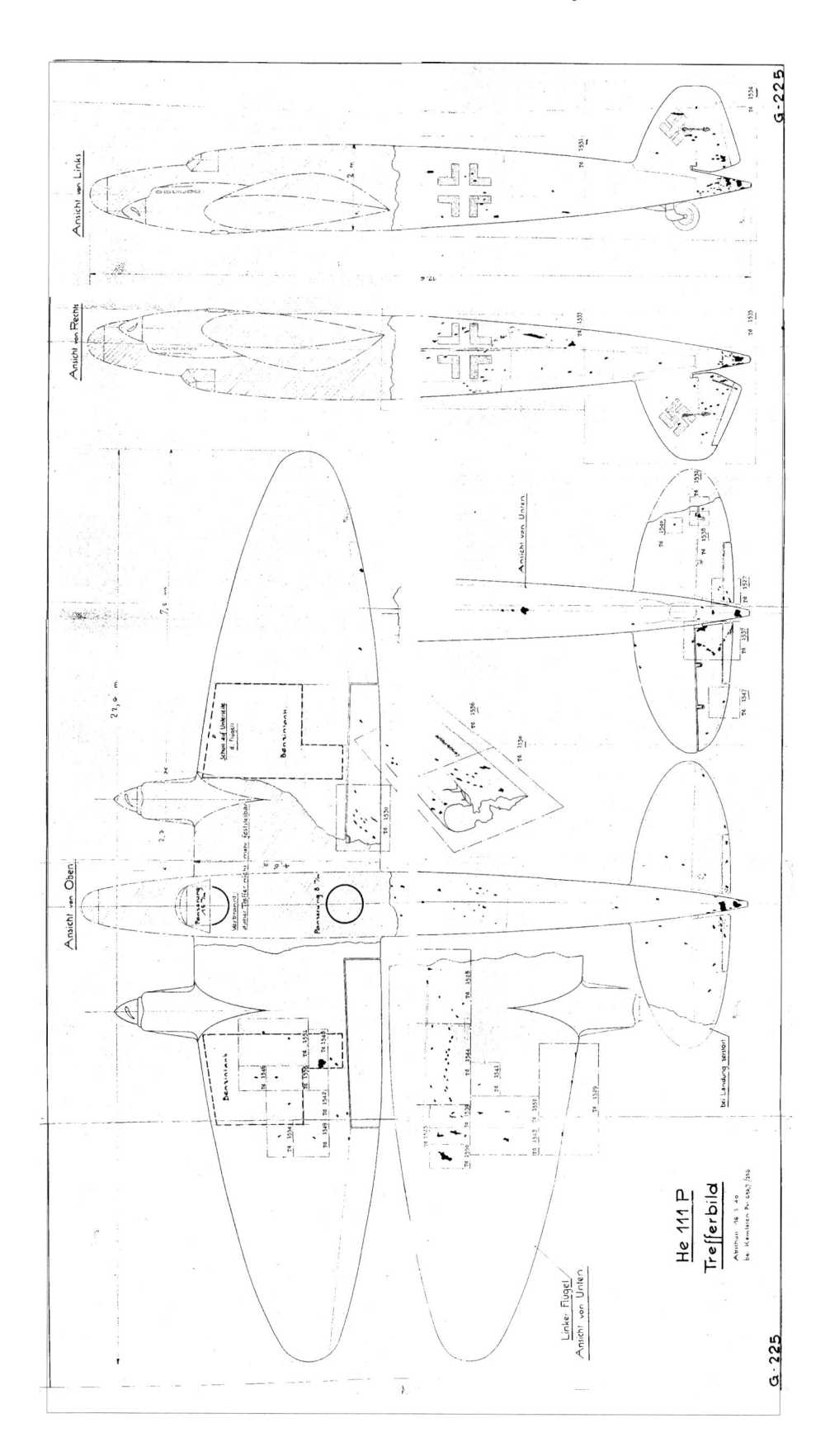

Appendix 18:
He 111 G1+HS, Ursins, June 2, 1940

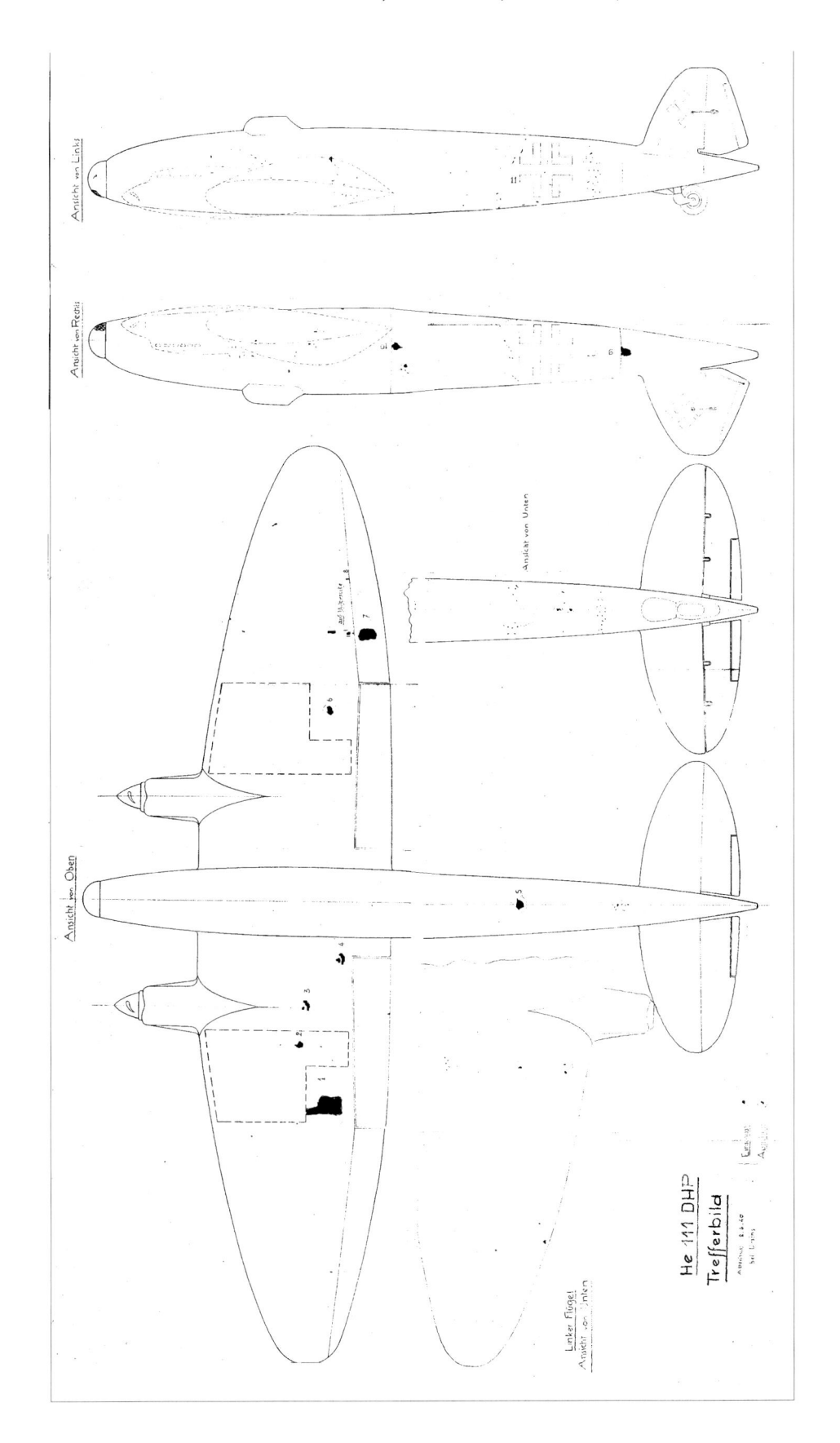

Appendix 19:
Avenches Airfield, Summer 1940
Fl. Abt. 3 (*Fl Kp 7/Fl Kp 8/Fl Kp 9*)

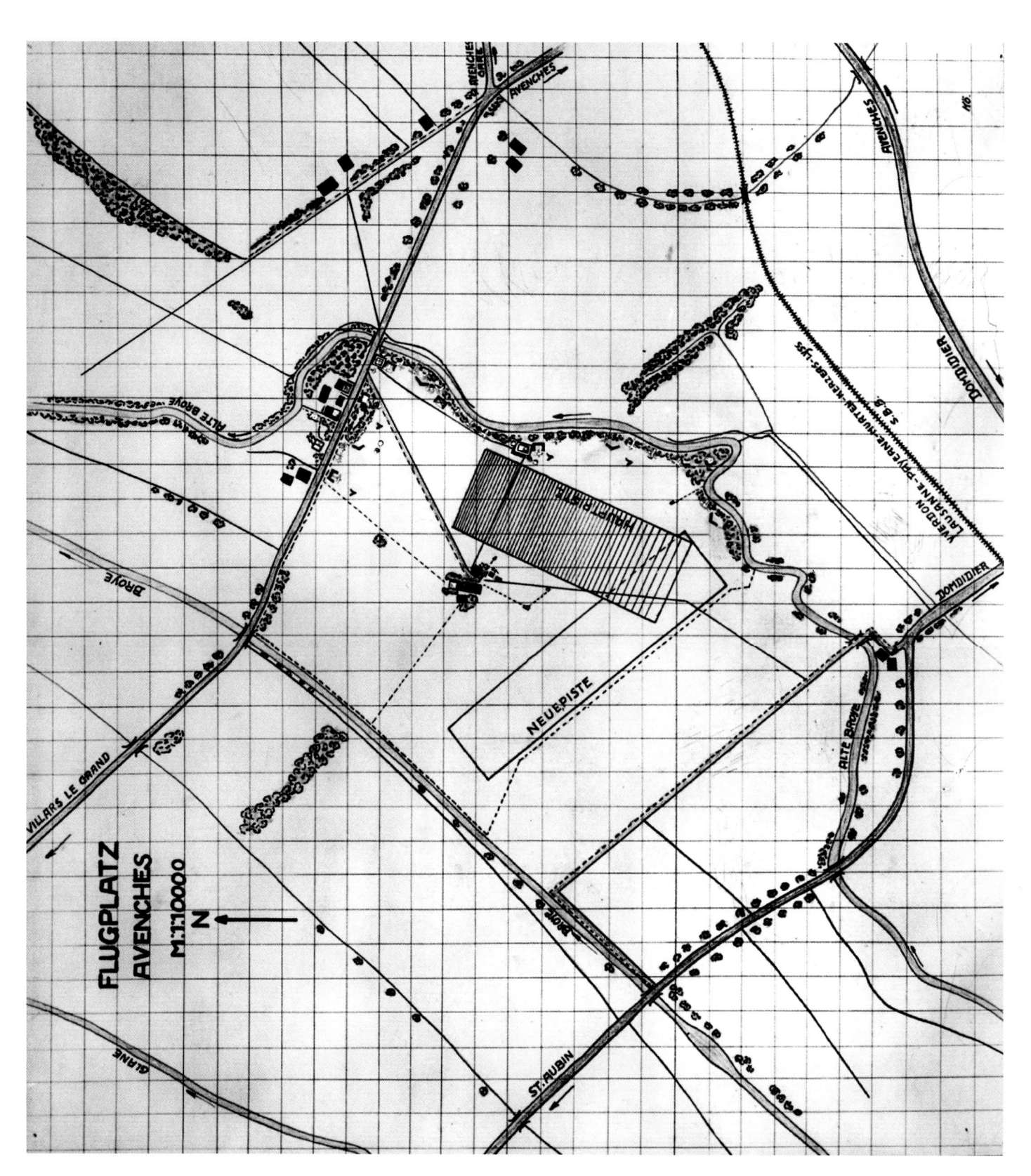

Appendix 20:
Interlaken Airfield, April 1942
Fl. Abt. 3 (*Fl Kp 7/Fl Kp 8/Fl Kp* 9)

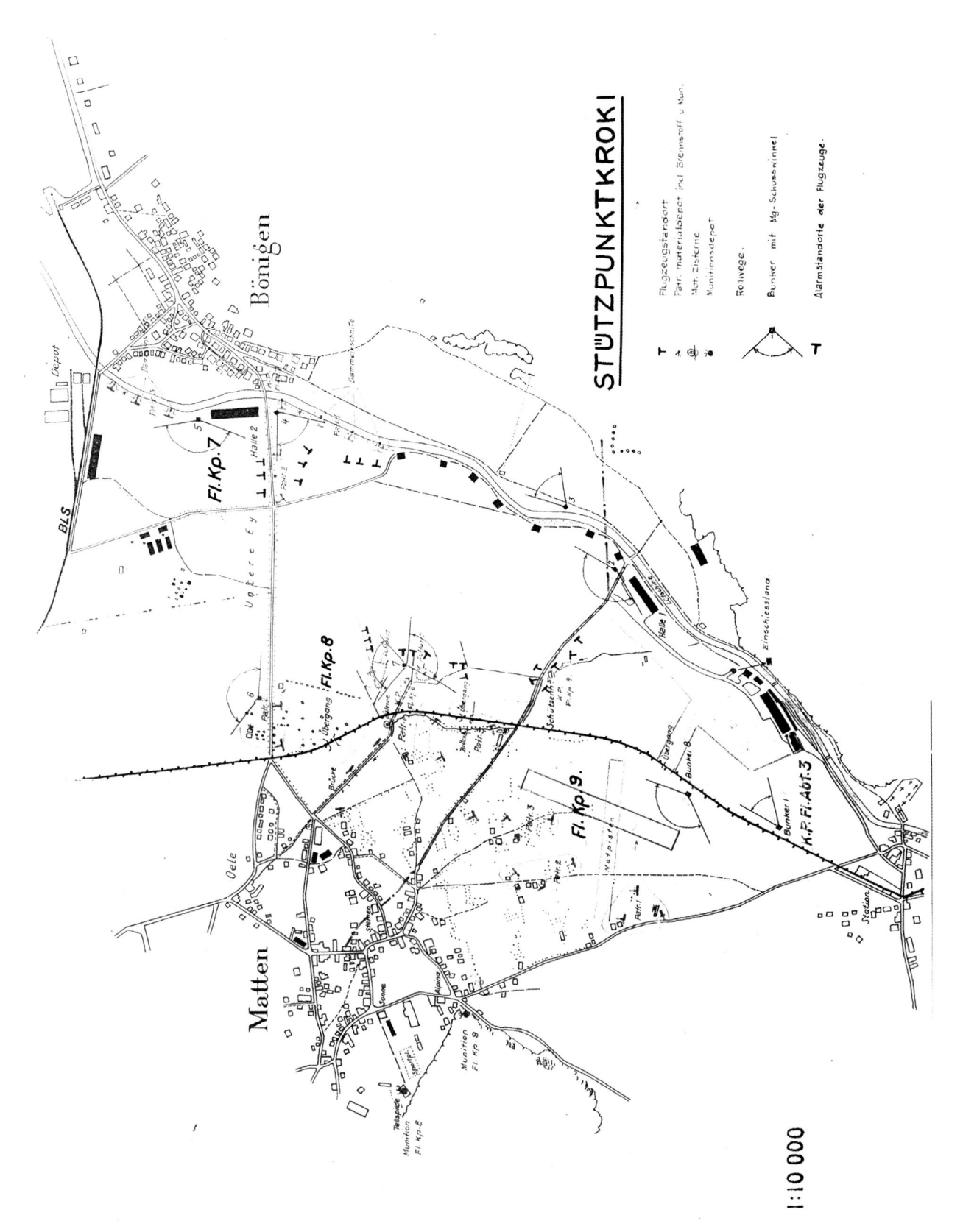

Appendix 21:
Zweisimmen Airfield, Cp av, December 1942

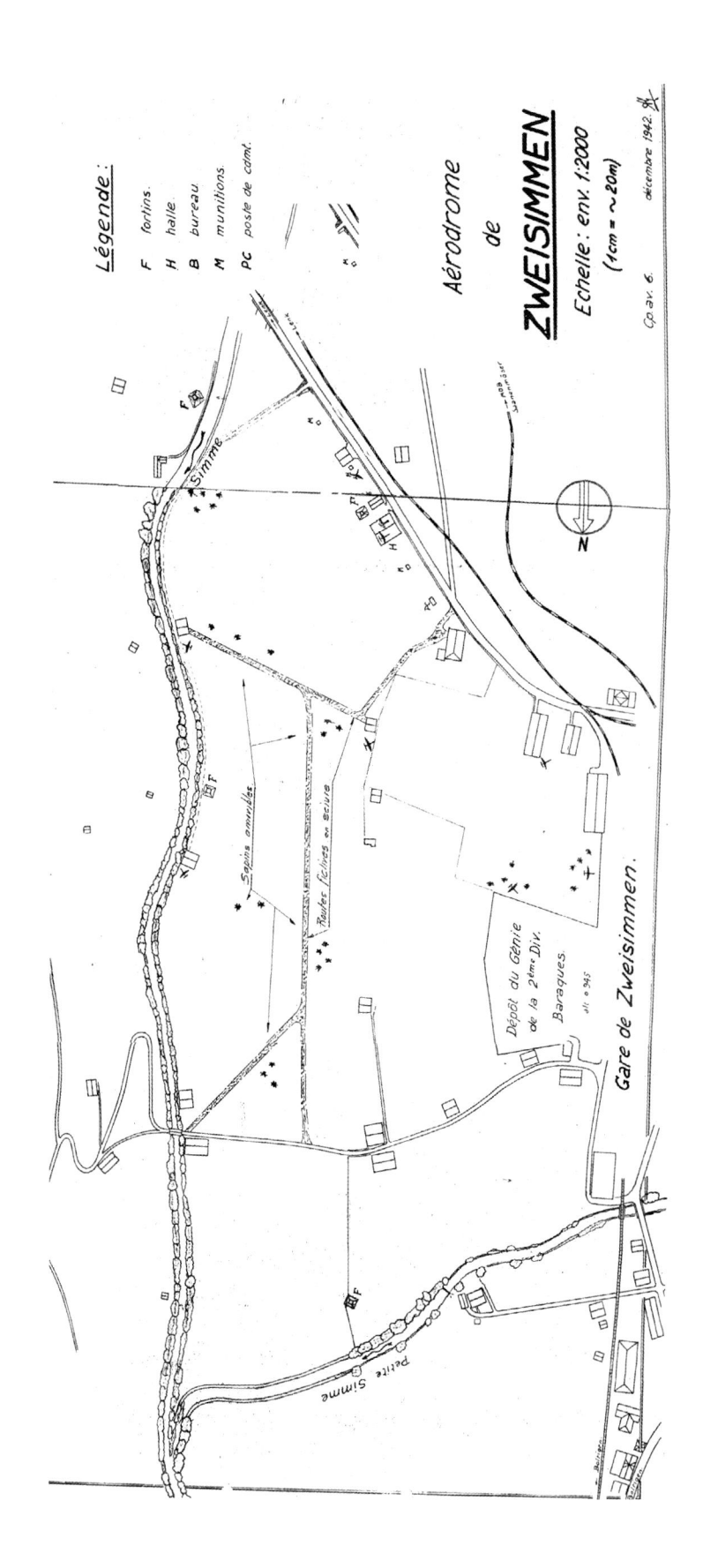

Appendix 22:
Unterbach Airfield, *Fl Kp* 9, July 1943

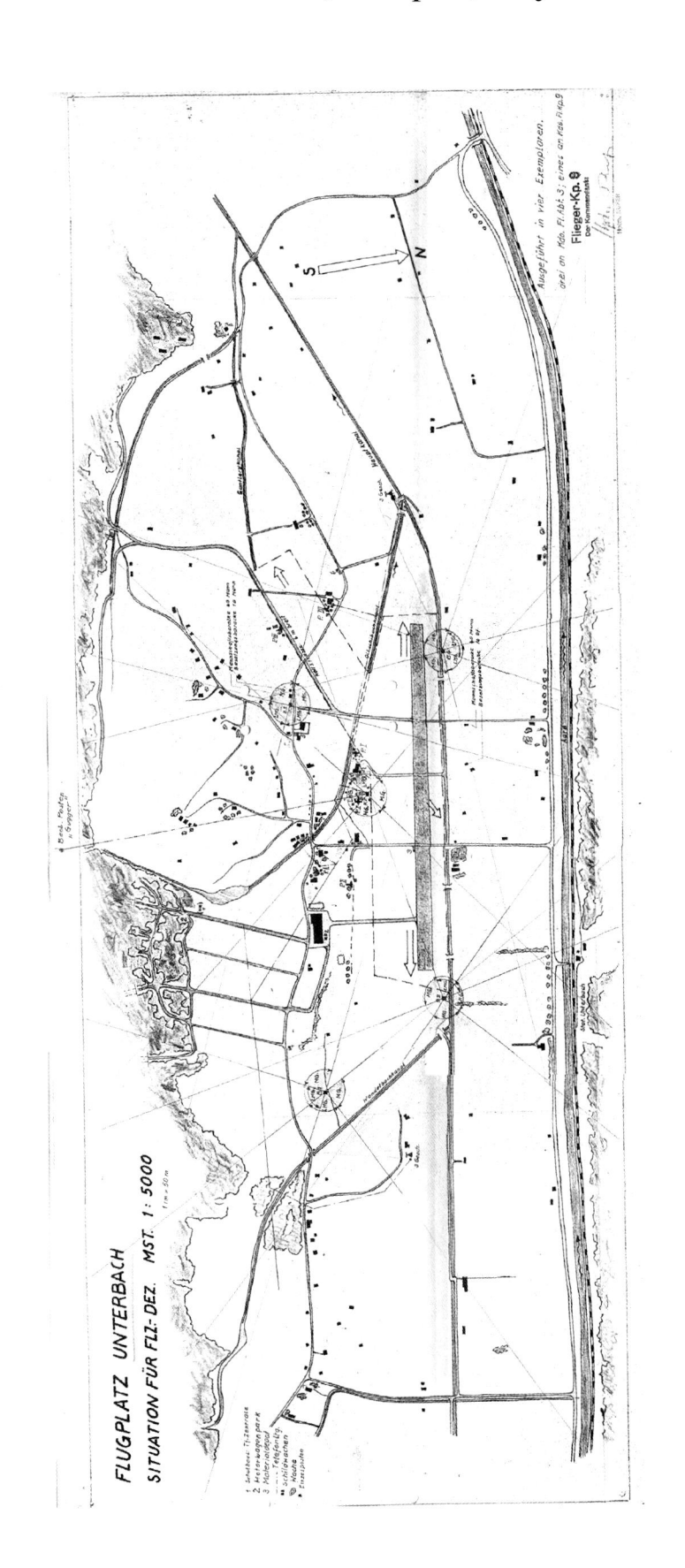

Messerschmitt Me 109 Units from 1939-1949

1939	**Régiment d'aviation 1** Cp av 6 Cap Roubaty	**Flieger Regiment 2** *Fl Kp* 15 *Hptm* Lindecker	**Flieger Regiment 3** *Fl Kp* 21 *Hptm* Hörning
1940	**Régiment d'aviation 1** Cp av 6 Cap Roubaty *Fl Kp* 7 *Hptm* Läderach *Fl Kp* 8 *Hptm* Fischer *Fl Kp* 9 *Hptm* Hitz	**Flieger Regiment 2** *Fl Kp* 15 *Hptm* Lindecker	**Flieger Regiment 3** *Fl Kp* 21 *Hptm* Hörning
1941	**Régiment d'aviation 1** Cp av 6 Cap Roubaty *Fl Kp* 7 *Hptm* Läderach *Fl Kp* 8 *Hptm* Fischer *Fl Kp* 9 *Hptm* Rufer	**Flieger Regiment 2** *Fl Kp* 15 *Hptm* Lindecker	**Flieger Regiment 3** *Fl Kp* 21 *Hptm* Hörning
1942	**Régiment d'aviation 1** Cp av 6 Cap Roubaty	**Flieger Regiment 2** *Fl Kp* 15 *Hptm* Lindecker	**Flieger Regiment 3** *Fl Kp* 21 *Hptm* Streiff
1943	**Régiment d'aviation 1** Cp av 6 Cap Liardon	**Flieger Regiment 2** *Fl Kp* 7 *Hptm* Läderach	**Flieger Regiment 3** *Fl Kp* 21 *Hptm* Streiff
1944	**Régiment d'aviation 1** Cp av 6 Cap Liardon	**Flieger Regiment 2** *Fl Kp* 7 *Hptm* Läderach	**Flieger Regiment 3** *Fl Kp* 21 *Hptm* Streiff
1945	**Régiment d'aviation 1** Cp av 6 Cap Liardon	**Flieger Regiment 2** *Fl Kp* 7 *Hptm* Läderach	**Flieger Regiment 3** *Fl Kp* 21 *Hptm* Streiff
1946	**Régiment d'aviation 1** Esc av 6 Cap Liardon	**Flieger Regiment 2** Fl St 7 *Hptm* Wiesendanger	
1947		**Flieger Regiment 2** Fl St 7 *Hptm* Wiesendanger Fl St 8 *Hptm* Brenneisen Fl St 9 *Hptm* Bizozerro Fl St 15 *Hptm* Suter	
1948-1949		**Flieger Regiment 2**	

Comments:

In 1943, the newly constructed *Fl Rgt* 4 and the UeG did not have any *Messerschmitt-Einheiten*.
In the course of the year 1947 Fl St 7, 9, and 15 retrained on the D 3801.
After the end of war in 1945 the *Flieger Kp* were newly organized. The flying personnel and the ground were from then on separated. The *Fliegereinheiten* were renamed to *Flieger-Staffeln*.

Messerschmitt Me 109 Flight Statistics

Number of Aircraft		Flying Hours	Flights
Aircraft on 12/31/38			
Me 109 D	2	---	---
Aircraft on 12/31/39			
Me 109 D	10	740.13	3171
Me 109 E	53	709.42	2868
Aircraft on 12/31/40			
Me 109 D	9	386.05	1194
Me 109 E	77	3059.06	10043
Aircraft on 12/31/41			
Me 109 D	9	182.42	812
Me 109 E	77	3073.52	9117
Aircraft on 12/31/42			
Me 109 D	9	256.30	923
Me 109 E	73	3459.52	8881
Aircraft on 12/31/43			
Me 109 D	9	216.29	864
Me 109 E	72	3165.17	9151
Aircraft on 12/31/44			
Me 109 D	9	269.44	976
Me 109 E	71	3617.49	9621
Me 109 G	12	205.17	703
Me 109 ?	0.29	1	
(no further information)			
Aircraft on 12/31/45			
Me 109 D	9	262.29	938
Me 109 E	75	2492.00	6859
Me 109 G	13	526.22	1058
Aircraft on 12/31/46			
Me 109 D	9	147.21	499
Me 109 E	66	2629.37	6045
Me 109 F	2	---	---
Me 109 G	11	491.30	1054
Aircraft on 12/31/47			
Me 109 D	8	96.57	321
Me 109 E	62	1549.48	2949
Aircraft on 12/31/48			
Me 109 D	7	75.27	227
Me 109 E	24	557.36	967
Aircraft on 12/31/49			
Me 109 D	0	65.03	153
Me 109 E	0	495.50	800

Messerschmitt Me 109 Dimensions and Performance Characteristics

(Further information found on page 13)

	Me 109 D-1	Me 109 E-3	Me 109 F-4	Me 109 G-6
Measurements:				
Wingspread	9900 mm	9900 mm	9924 mm	9924 mm
Length	8700 mm	8800 mm	8940 mm	8940 mm
Weight:				
Total weight	1505 kg	1855 kg	2093 kg	2330 kg
Loading capacity	493 kg	730 kg	1070 kg	1070 kg
Max. takeoff weight	1998 kg	2585 kg	3263 kg	3400 kg
Performance:				
V. max. horizontal	470 km/h	570 km/h	600 km/h	635 km/h
Max. climbing capacity	12.0 m/s	14 m/s	14.8 m/s	15.6 m/s
Ceiling	8000 m ü M	10 100 m ü M	9000 m ü M	9000 m ü M
Flight duration	1 h 30 min	1 h 45 min	1 hr 30 min	1 hr 30 min
Range	600 km	600 km	600 km	700 km
Engine	Jumo 210 D	DB 601 Aa	DB 601 E	DB 605 A-1
Takeoff power	680 PS	1100 PS	1250 PS	1450 PS
Revolutions	2700 t/min	2400 t/min	2700 t/min	2800 t/min
Piston displacement	19.7 lt	33.0 lt	33.0 lt	35.7 lt
Type	V12 carburetor	V12 injection	V12 injection	V12 injection
Weight	442 kg	610 kg	685 kg	720 kg
Propeller	VDM ø 3200 mm	VDM ø 3100 mm	VDM 9.12010a ø 3000 mm	VDM 9-12087 ø 3000 mm
Firearms	4 MG 2900 2 FF-K	2 MG 29 1 MG 151/20	2 MG 17 1 MG 151/20	2 MG 131
Dropping weapons	--	2 50 kg Bb	None in Switzerland	None in Switzerland
Radio equipment	FuG VII	FG IX (SIF 450)	FuG VIIa + FuG25	FuG 16Z
Number of aircraft	10	89	2	14
Aircraft numbers	J 301 to J-310	J-311 to J-399	J-715 and J-716	J-701 to J-714

Comments: Concerning the exact measurements of the Me 109, there are contradictions present in the literature. Even in official documents there are differing statements, or the measurement is given with "~."
The statements on the flight performance are guidelines. The values do often not agree according to manufacturer, technical handbooks, or documents.

The dates come from the following sources:
Me 109D = Operating manual BF 109 C&D
Me 109E = Measurement certificate of Messerschmitt AG from August 1941
Me 109G = Operating instructions L. Dv.T.2109 G-2, G-4, G-6/Fl from June 1943
K. Urech, *Die Flugzeuge der schweizerischen Fliegertruppe seit 1914*

Messerschmitt Me 109 Dates

Aircraft No.	Factory No.	Purchase	Written off	Aircraft Hours	Engine No.	Incident Date	Incident	Comment	Location	Pilot	Fl Kp/St
J-301	2297	1/19/39		41562			Acquisition				
				41562		5/31/39	Emergency landing	Engine failure (Junker guarantee)	Biel	Thurnheer H	
				41695		8/22/40	Aborted takeoff	Engine failure, fuel supply	Thun	Huser P	
				41691		11/23/43	Emergency landing	Engine failure, supercharger defect	Dübendorf	Hofer W	7
			12/28/49	41647		2/28/44	Written off	Engine failure, supercharger shaft eroded, engine written off			
J-302	2299	1/19/39		41576			Acquisition				
						3/3/39	Air service	Tip damaged	Dübendorf	Kisling R	21
						4/4/39	Air service	Tip and landing gear damaged	Dübendorf	Wüthrich M	
						5/26/39	Air service	Tip damaged	Thun	Schäfter A	15
						3/15/40	Air service	Wing deformed	Payerne	Rufer A	UeG
				41893		5/11/45	Emergency landing	Engine failure, wrong fuel, 90 OZ instead of 93 OZ	Thun	Kolb F	
						12/19/45	Landing failure	Left wing grazed ground	Payerne	Eggenberger M	9
				41573			Engine change				
			12/28/49	41893			Written off				
J-303	2295	12/17/38		41552			Acquisition				
					12/17/38		Landing failure	Excessive speed, aircraft repairs at B.F.W	Dübendorf	Mooser E	KTA
					2/18/39		Emergency landing	Engine failure	Dübendorf	Wild J	A Fl Pk
				41906	5/12/40		Emergency landing	Engine failure, supercharger defect	Lodrino		
			2/9/48	259.18	7/27/46		Crash landing	ev caused by Böe, total loss	Meiringen	Bachofner H	Kdt Rgt 4
J-304	2298	1/10/39			41573		Acquisition				
						2/26/47	Trimming	Stabilo adjustment, chain brakes too weak	Dübendorf	Wüthrich M	
					41876		Engine change				
			12/28/49		41906		Written off				
J-305	2300	1/5/39			41647		Acquisition				
						1/5/39	Emergency landing	Engine failure, coolant loss, repairs at BFW Ellikon ZH	Wyss E	KTA	
						2/29/40	Emergency landing	Engine failure	Payerne	Homberger R	15
			12/28/49		41552	2/28/43	Written off	Engine failure	Lodrino	Läderach W	7
J-306	2301	1/10/39			41657		Acquisition				
15						4/2/42	Emergency landing	Engine failure, broken oil pipe	Grenchen		Egli F
				236.29	41904		Engine change				
					41909	8/4/42	Emergency landing new fuselage	Magnetic drive coupling defect,	Dübendorf	Schneider P	9
					41647		Engine change				
			12/28/49		41695		Written off				

Aircraft No.	Factory No.	Purchase	Written off	Aircraft Hours	Engine No.	Incident Date	Incident	Comment	Location	Pilot	Fl Kp/St
J-307	2302	1/7/39			41646	2/1/39 6/10/39 4/26/47	Acqisition Belly landing Landing failure Landing gear Written off	Landing gear defect Supercharger and landing gear damaged Left shock strut buckled while taxiing	Dübendorf Payerne Dübendorf	Läderach W Boudry L Genner F	15 (2) UeG
		12/28/49			*41904*						
J-308	2303	1/10/39			41691 41576	12/15/39 11/16/43 1/4/46 9/2/47	Acquisition Emergency landing Emergency landing Landing failure Crash landing	Break in fuel supply line Engine failure, eroded Tail wheel column damaged Total loss	Dübendorf Dübendorf Dübendorf Payerne	Streiff V Ernst E Kössler E	21 8 UeG Näf U
9			9/2/47	226.25	41573		Written off				
J-309	2304	*1/19/39*				1/17/39 4/1/39 11/22/45	Acquisition Emergency landing Landing failure Emergency landing	Other acquisition date 1/17/39 Foreign matter in oil pump Tip slightly damaged Engine failure, spring break on inlet valve no. 7	Dübendorf Dübendorf Agno	Frei W Köpfli, M Schwärzler G	21 UeG
			12/28/49				Written off				
J-310	2305	1/5/39		42.11 69	41696	4/20/39	Acquisition Landing failure Air combat	Due to repairs in Augsburg, factory flight on 2/17/39 Left wing damaged, Mtt for repairs Shot down by Me 110 of the ZG 1	Dübendorf Boécourt	Boudry L Rickenbacher R(t)	(2) 15
J-311	2159	7/5/39	6/4/40	7.21	10640	11/24/39 1/12/40 4/15/42	Acquisition Emergency landing Emergency landing Collision	Coolant loss, pipe burst Engine failure, fuel pump lever break During air combat practice wing grazed J-359	Dübendorf Dübendorf Thun	Streiff V Wannenmacher E Reber A	21 21 15
Pilatus Stans			12/2/49	161.48	10992 *11006*	11/17/42 3/5/45	Parts revision Normalization Written off				
J-312	2160	6/20/39	3/3/40	4.26 47.25	10647		Acquisition Crash	Total loss, Stabilo chain brakes released	Ottikon	Streiff V	21
J-313	2161	6/20/39		3.04	10643 10995	8/22/39 1/24/45	Acquisition Collision Normalization	Taxiied into coal shed at landing	Dübendorf	Giger H	A Fl Pk
Pilatus Stans			10/1/47	364.35	10946		Decommissioned	Written off on 8/14/48			
J-314	2162	5/7/39 6/7/39	106.46	10646			Acquisition Crash	Prototype aircraft for weapons installation Loss of speed	Doflug Altenrhein	Suter G (t)	KTA

Aircraft No.	Factory No.	Purchase	Written off	Aircraft Hours	Engine No.	Incident Date	Incident	Comment	Location	Pilot	Fl Kp/St
J-315	2163	6/20/39		3.36	10661 10962	7/26/43 3/15/45	Acquisition Engine change Taxi maneuver Normalization	Taxied into hall gate, brake defect	Unterbach	Tech Of	9
Pilatus Stans			11/20/46	372.46	40698		Written off				
J-316	2164	6/20/39	6/26/42	2.39 187.07	10664 10905 10664		Acquisition Engine change Crash	Collision on mountainside	Schallenberg	Nipkow G (t)	9
J-317	2165	6/23/39		7.39 49.37	10697 10998	6/29/39 7/11/39 3/13/40 6/29/40	Acquisition Emergency landing Propeller Emergency landing	Engine failure, foreign substance in oil pump St takeoff control pushed over, propeller defect Engine failure, eroded, engine change Canopy lost during flight	Dübendorf Altenrhein Geneva	Frei W Meyner H Fischer A	A Fl Pk 21 8
		Widmer E	8		10749	11/2/44	Normalization				
Pilatus Stans			12/28/49		10994	6/17/47	Taxiing accident	Taxied into fuel transport, misunderstanding with *Rollwart*	Thun	Zschokke V	8
J-318	2166	6/28/39		186.19 271.32	10691	5/1/40 8/20/42 4/20/44	Acquisition	Engine overheated on ground, loss of water Payerne Wing deformed in flight Left wing grazed web plate and fuselage cowling	Buochs		8 Det. 53
Pilatus Stans					10993	11/2/44	Normalization				
			8/14/48	402.32	10993 10992	2/6/45 10/1/47	Taxi accident Written off	Left landing gear buckled Aircraft and engine decommissioned	Dübendorf	Feldmann F	7
J-319	2167	6/23/39		4.14	10695 *10698* 10768	8/16/44 10/10/44	Acquisition Emergency Landing Normalization	Engine failure, engine cutout	Interlaken	Widmer E	A Fl Pk
Pilatus Stans			8/14/48	411.03	10947	3/28/47 10/1/47	Tire blow-out Written off	Cheval de bois at landing Aircraft and Engine decommissioned	Payerne	Zschokke V	8
J-320	2168	6/24/39		3.03 113.49 7	10686 10686	3/21/40 6/9/42	Acquisition Engine failure	Engine failure, eroded Landing failure			
Buochs		Kasser H				3/15/45	Normalization	Belly landing 50% damage			
Pilatus			5/9/47	154.26	10768		Crash	O_2 deficiency during high altitude flight	Anenfirn VS	Aschwanden E (t)	9

Aircraft No.	Factory No.	Purchase	Written off	Aircraft Hours	Engine No.	Incident Date	Incident	Comment	Location	Pilot	Fl Kp/St
J-321	2169	6/28/39		3.49	10698 10763 10982	9/6/39 9/15/44 10/18/44	Acquisition Engine change Engine change Emergency landing Normalization Takeoff failure	Engine failure, oil tank torn Propeller adjustment, taxied into Lütschine	Altenrhein Pilatus Stans Interlaken	Hörning W Ernst F	21 15
			12/28/49			1/27/47 4/3/47	Emergency landing Landing failure Written off	Engine failure, hole in piston No. 6 re Aircraft swerved, minor damage	Sion Dübendorf	Arn F Rageth E	15 7
J-322	2170	6/24/39	4/24/40	3.01 56.56	10696	7/14/39	Acquisition Crash Written off	Engine failure, pilot severely injured, aircraft completely destroyed	Mollis	Wannenmacher E	21
J-323	2171	6/27/39	8/14/48	2.22 10.34 138.07 369.37	10702 10768 10694	1/10/40 6/1/42 10/1/47	Acquisition Engine failure Engine change Decommissioned Written off	Engine failure, eroded Engine change, TM 1016 Aircraft and engine decommissioned	Dübendorf Buochs		
J-324	2172	6/27/39		3.0	10701 11002	1/9/40 1/10/41	Acquisition Engine change Chute landing	Fuselage rammed Tire blow-out at takeoff	Dübendorf		
Dübendorf		Schärer R				12/3/42 7/14/43 8/28/43 9/5/44	Landing failure Takeoff failure Canopy lost in flight Emergency landing	Aircraft veered off Ground grazed with left wing Belly landing after shots from USAAF P-51	Sion Dübendorf Sempachersee Dübendorf	de Meyer Rieser Rieser J Heiniger R (verl)	UeG UeG 7
Pilatus Stans			12/28/49		10984	4/30/45	Normalization Written off				
J-325	2173	6/28/39		2.47	10704 63259 10773 63259 10706	5/1/40 5/11/44 11/9/44 4/12/45	Acquisition Engine change Crash landing Engine change Emergency landing Crash landing Normalization	Loss of speed, aircraft heavily damaged Engine failure, manifold pressure regulator Heavy gust	Thun Forel Emmen	Benoit F Dannecker F Mühlemann E	6 8 21
Pilatus Stans											
Buochs		Dosch J	12/28/49	8		6/7/45	Written off	Tire blow-out at takeoff			

240

Aircraft No.	Factory No.	Purchase	Written off	Aircraft Hours	Engine No.	Incident Date	Incident	Comment	Location	Pilot	Fl Kp/St
J-326	2174	6/27/39			10705	4/17/40	Acquisition		Lausanne	Schenk P	6
						1/16/43	Emergency landing	Connection to cylinder shutoff broken Canopy lost in flight			
Reg. Saanen		Leuenberger A		9	11045	4/11/45	Engine change	Takeoff failure with subsequent belly landing, propeller at wrong gradient	Thun	Reber A	15
					10697	1/22/48	Engine change	Tire blow-out at landing	Payerne	Brenneisen M	8
			12/28/49		10697		Written off				
J-327	2175	6/27/39		1.50	10694		Acquisition				
				22.51	11035	2/29/40	Belly landing	Landing gear, oil pipe break	Dübendorf	Roubaty J	6
				24.47	10643	4/20/40	Takeoff failure	Overturned, aircraft severely damaged (in repairs until June 42)	Dübendorf	Ludwig T	
				400.04	10973	3/9/45	Normalization				
F+W Emmen			8/14/48	218.02	10765		Written off				
J-328	2176	6/27/39		2.16	10693		Acquisition				
				74.55		6/8/40	Emergency landing	After bombardment in air combat	Biel-Bözingen	Homberger R (verl)	15
					10996	3/10/41	Landing failure	Nose-over	Buochs	Soldenhoff W	
			8/14/48		10757	3/9/45	Normalization	Written off	Pilatus Stans		
J-329	2177	6/28/39		4.08	10700		Acquisition				
				21.18	10991		Engine change				
				129.27		6/4/40	Emergency landing	Engine eroded After bombardment in air combat	Biel-Bözingen	Rufer A	15
Buochs				362.07	10643	12/21/44	Parts revision				
					10701	8/14/48	Normalization Written off		Pilatus Stans		
J-330	2178	6/29/39			10706	7/16/40	Acquisition Overturned		Avenches	Benoit JP	6 (7)
					10679	12/3/42	Engine change	Landing	Interlaken		
					10997	3/4/44	Engine change	Throttle control defect	Interlaken		
					20113	2/2/45	Normalization	Installation of oxygen system and bomb assembly	F+W		
			8/14/48	417.22	10752			Written off			
J-331	2189	6/29/39			10757		Acquisition		Dübendorf	Trenkler	Mtt
				159.32	10967	2/15/41	Emergency landing	Overflight Augsburg – Düb, aircraft swerved at landing	Buochs	Meyner H	21
				202.52	11081	11/15/43	Parts revision	Fuel pump	Buochs		
						2/3/45	Takeoff failure	Aircraft swerved	Payerne	Dumont A	
						2/20/45	Normalization	Installation of oxygen system and bomb assembly	Pilatus		
						4/13/46	Emergency landing	Fuel pump	Emmen	Rufer A	9

241

Aircraft No.	Factory No.	Purchase	Written off	Aircraft Hours	Engine No.	Incident Date	Incident	Comment	Location	Pilot	Fl Kp/St
						5/23/46	Collision	Taxied into parked AC-12 HB-OSI	Lausanne	Liardon F	6
			8/14/46				Written off				
J-332	2190	6/29/39		269.18	10968						
				1.51	10764	12/8/39	Emergency landing	Engine failure	Dübendorf	Ettinger A	8
						3/8/40	Emergency landing	Engine failure	Geneva	Brenneisen M	8
						4/2/40	Emergency landing	Engine failure	Thun	Brenneisen M	
			12/23/46	339.48	10686		Written off	Poor condition, different date 11/20/46			
J-333	2191	6/29/39					Acquisition				
				2.57	10752	8/31/39	Emergency landing	Propeller adjustment, heavy damage	Zimikon	Köpfli M	21
				6.25		4/14/42	Emergency landing	Engine failure, overturned	Biel-Bözingen Arn F		15
						10/24/44	Normalization		F+W Emmen		
			8/14/48	276.52	10064		Engine change	Engine number 10700 / 10957			
							Written off				
J-334	2192	6/29/39					Acquisition				
				1.51	10772	8/9/39	Emergency landing	Propeller adjustment	Schwerzenbac	Ettinger A	h
						4/29/40		Tail wheel dented at landing	Payerne	Künzler F	7
						10/30/42		Collision with high voltage cable	Wimmis	Siegfried H	7
					10752			Engine change			
				187.06	10976	7/7/44	Crash	Wing break	Kägiswil	Brenzikofer F (t)	9
J-335	2193	6/30/39					Acquisition				
				1.29	10773		Engine change				
				46.51	10913	5/1/40	Emergency landing	Loss of water at cylinder 1	Payerne		7
				224.35	10993	11/11/43		Engine failure, connecting rod break	Interlaken	Fischer A	8
					10760	4/30/45	Normalization				
Pilatus Stans			8/14/48	398.16	10979		Written off				
J-336	2194	6/30/39					Acquisition				
				2.22	10756	3/19/40	Emergency landing	Engine failure, mix regulator	Dübendorf	Meyner H	21
						6/20/41		Canopy lost during flight	Buochs	Aschwander E	15
						7/28/41	Landing failure	Fuselage rammed	Avenches	Reber	?
						2/15/43	Landing failure	Fuselage rammed	Emmen		
						4/10/43	Emergency landing	Engine failure, carburetor nozzle clogged	Emmen	Köpfli M	21
						4/20/43		Wing deformation, overstress in air combat practice	Emmen	Köpfli M	21
						7/5/44	Emergency landing	Engine failure, fuel supply	Kägiswil	Egli ?	
						11/20/44		Normalization			
Pilatus Stans			8/14/48	319.57	11888		Written off				
J-337	2195	6/30/39					Acquisition				
				2.16	10760	8/27/40	Emergency landing	Engine failure, eroded	Emmen	Wannenmacher E	21
				27.31	10956	1/19/45	Normalization				
Pilatus Stans			11/20/46								
					315.27	4/8/46	Emergency landing	Engine failure, oil filter silted up	Unterbach	Bridel A	
							Written off				

242

Aircraft No.	Factory No.	Purchase	Written off	Aircraft Hours	Engine No.	Incident Date	Incident	Comment	Location	Pilot	Fl Kp/St
J-338	2196	6/30/39		2.06	10767		Acquisition		Interlaken	Ernst E	8
					10774	9/15/44		Caught fire during taxiing, leaky fuel line	F+W Emmen		
					10906	11/11/44	Normalization	Landing gear defect, wrong hydraulic oil	Payerne	Burlet G	6 Rgt 4
						12/23/44	Belly landing	Engine failure, sand in manifold	Thun	Asper E	Kdt Rgt 4
			8/14/48	310.29	10906 10661	3/13/46	Belly landing Written off	pressure regulator casing			
J-339	2197	6/30/39		2.26	10769		Acquisition		?	Köpfli M	21
					10700	9/30/39	Chute landing	Landing gear damaged			
						1/19/45	Normalization				
Pilatus Stans			10/2/46		10749		Crash	Collision with J-344	Raron	Kilchenmann P (t) 9	
J-340	2198	6/30/39		2.35	10759	3/9/40	Acquisition	Engine failure, eroded	*Emmen*	Ground crew	21
				27.47	*10995*	5/24/40	Engine change	During taxiing landing gear buckled	Buochs	Boudry L	
						3/16/42	Landing failure	Swerved during landing, tail wheel wobbling			
					10971	3/4/44	Crash landing	Loop in snow	Interlaken	Schmid F	7
						2/2/45	Normalization				
Pilatus Stans			12/28/49			1/20/48	Written off	Tire blow-out, swerved at landing	Payerne	Zschokke V	8
J-341	2362	10/1039		2.32	10905	4/27/40	Acquisition	Cheval de bois, severe damage, repairs until 10/15/41	Dübendorf	Willi G	21
				45.47	10905		Takeoff failure	Wing change due to deformation			
				89.58	10982	2/10/44	Normalization		Interlaken		
			8/14/48	244.13	10917	9/27/44	Written off		F+W Emmen		
J-342	2363	10/8/39		3.03	10906		Acquisition		Payerne	Boudry L	
				31.06	10702	7/29/40		By Cheval de bois shock strut casing damage			
				31.15	*10752*	?	Emergency landing	Engine failure, supercharger defect	Mollis	Bider ?	
					10702	3/4/43	Emergency landing		Dübendorf	Stutz A	UeG
						9/18/44	Normalization				
J-343	2364	10/8/39	8/14/48	341.11	11002		Written off		Dübendorf		
				2.53	10904	1/4/40	Acquisition	Engine failure, eroded			7
				7.32	10695	1940	Engine change	Engine change			8
				14.39	10912	4/12/40	Engine change	Explosion left FF-K, wing change	F+W Emmen		(19)
			8/14/48	335.57	10912	9/14/44	Normalization	after 200 m wing grazed ground	Dübendorf	Moll A	
						7/3/47	Takeoff failure Written off				
J-344	2385	10/10/39		2.34	10948		Acquisition				9
				70.31	*11035*	7/20/41	Engine and propeller change (operating error ?)		Thun		
					10693	8/27/41	Engine change, control shaft drive defect				

Aircraft No.	Factory No.	Purchase	Written off	Aircraft Hours	Engine No.	Incident Date	Incident	Comment	Location	Pilot	Fl Kp/St
					10765	11/11/44	Normalization		F+W Emmen		
				445.32	10640	4/11/45		Collision with drag chute during shooting flight	Payerne	Wiesendanger F	7
			10/2/46					Collision with J-339	Raron	Vivian E (t)	9
J-345	2386	10/10/39		2.46	10903	4/2/40	Acquisition Emergency landing	Water line break	Payerne	Künzler F	7
					10765	4/8/40	Engine change				
					11049	11/22/43	Belly landing	Landing gear defect	Thun	Liardon F	6
					10901	1/16/45	Normalization				
Pilatus Stans			7/7/47	405.53	30837		Emergency landing	Engine failure, tear in line to oil cooler, aircraft total	Fribourg	de Pourtales L	
J-346	2403	10/8/39		2.03	10946	11/29/40	Acquisition Landing failure	Landing gear not completely released	Thun	Widmer E	8
						11/20/41	Chute landing	Fuselage rammed		Köpfli M	21
				79.53	10704	3/20/44	Engine change	Test with Swiss piston rings			
					10998	9/29/44	Engine change	Installation of EW propeller			
					10686	1/29/45	Normalization	Oil line defect	Thun	Ernst F	15
F+W Emmen			8/14/48	299.02		11/21/46	Emergency landing Written off	Propeller adjustment	Zweisimmen	Suter R	15
J347	2404	10/10/39		1.42	10912	11/11/41	Acquisition	Test with EW propeller			
						9/17/42		Installation of EW propeller			
			12/23/46	324.29	10756	1/15/45	Normalization	Written off	F+W Emmen		
J-348	2405	10/12/39		1.55	10923	4/30/41	Acquisition Engine change	Engine failure, piston ring broken	Thun	Widmer E	A Fl Pk
					10694	1/16/43	*Emergency landing*	Engine failure	Interlaken		
					10904	10/2/44	Normalization		F+W Emmen		
			12/23/46	432.19	10901		Written off				
J-349	2406	10/26/39		1.36	10917	5/16/40	Acquisition Air combat	Bullet holes	Dübendorf	Streiff V	21
					10768	11/25/44	Engine change Normalization		F+W Emmen		
			2/8/45	201.19	10998		Crash	Poor weather, parachute jump	Erwil	Ernst F	15
					10998						
J-350	2407	10/26/39		3.19	10907	2/29/40	Acquisition Engine change	Engine failure, eroded	Buochs	Borner A	21
				20.32	10697	4/3/44	Landing failure	Swerved, nose-over	F+W Emmen		
			10/1/46	365.35	10979	10/3/44	Normalization	Emergency landing after wing deformation from bomb release	Locarno	Rosenmund P	8

244

Aircraft No.	Factory No.	Purchase	Written off	Aircraft Hours	Engine No.	Incident Date	Incident	Comment	Location	Pilot	Fl Kp/St
J-351	2408	10/26/39		2.59	10982	12/28/39	Acquisition	Tail wheel damaged on frozen ground	Dübendorf	Bouvier E	
					10997	4/10/40	Engine change	Engine failure, fan broken	Payerne		
						8/26/40	Parking damage	rammed by CV-E (C-802). 4600 francs damage			
					10997	3/31/42	Emergency landing	Lack of fuel	Reichenbach	Aschwanden E	15
					10995		Engine change				
						2/10/44	Canopy lost during flight				
					10694	12/2/44	Normalization				
			8/14/48		10760		Written off		F+W Emmen		
J-352	2409	11/7/39		5.16	10983		Acquisition		Lausanne	Hadorn E	6
						6/19/42	Landing failure	Wing grazed ground	Belp	Meyner H	21
						4/4/44	Aborted takeoff	Engine failure			
					10095	9/26/44	Normalization				
F+W Emmen			8/14/48	308.10	11086		Written off				
J-353	2420	10/26/39		2.08	10986	11/28/39	Acquisition		Dübendorf	Wachter A	6
							Tip and aileron damaged				
					10903	11/15/41	Engine change				
					10763	4/5/45	Normalization		F+W Emmen		
			6/27/46	327.12	10763		Emergency landing broken, aircraft total	Propeller adjustment, regulator rod	Yverdon	Arn F (verl)	15
J-354	2421	11/7/39		2.30	11005		Acquisition		Isenfluh BE	Fleury M	6
				208.30	10759	12/2/41	Engine change				
			10/12/43		10759		Crash	Lack of fuel, parachute jump			
J-355	2422	11/7/39		3.31	10979	2/13/40	Acquisition		Dübendorf	Frei W	9
				16.09		5/25/40	Emergency landing	Engine failure	?		
							Chute landing	Fuselage rammed left, engine bearer displaced			
				64.15	10984	7/28/42	Engine change		Interlaken		
				166.10	30080	10/11/44	Engine change		Buochs		
				177.03	30080	1/10/45	Normalization		Pilatus Stans		
				240.17	10901	11/13/46	Engine change		Buochs		
				312.30	31291	9/8/49	Engine change		Buochs		
				323.01	31291	12/28/49	Written off				
						6/30/59	Lucerne Museum of Transport				
						6/1/79	Dübendorf Fliegermuseum				
J-356	2423	12/16/39		3.19	11088	2/12/45	Acquisition		F+W Emmen	Ernst E	8
					10769	3/11/45	Normalization	Impeller rim broken	Dübendorf		
			8/14/48	401.24	10906		Landing				
							Written off				
J-357	2438	4/16/40		3.03	10973	5/14/40	Acquisition		?	Rieser ?	7
					10973	8/19/42	Crash landing	Tail wheel fittings torn	Kloten	Lutz P (verl)	
					10767	12/23/44	Normalization	Aircraft severely damaged, repairs until 8/28/43			
			8/14/48	238.11	11081		Written off		F+W Emmen		

Aircraft No.	Factory No.	Purchase	Written off	Aircraft Hours	Engine No.	Incident Date	Incident	Comment	Location	Pilot	Fl Kp/St
J-358	2425	12/15/39		3.00	11055	4/22/40	Acquisition		Payerne	Schwegler P	
				25.06		5/26/40	Landing failure	Cheval de bois, Landing gear defect			
						8/11/42	Chute landing	Fuselage rammed left	Buochs	Schwarzenbach R	8
					11037	4/3/44	Crash landing	Stalled at landing, medial damage	Buochs	Borner W	21
					10704	10/23/44	Takeoff failure	Nose-over, 2000 francs damage	Uetendorf	Bueche J	6
						2/15/45	Belly landing	Engine failure, 42,000 francs damage	F+W Emmen		
			8/14/48	304.17	10704	4/18/47	Normalization	Cheval de bois	Payerne	Dumont A	
							Landing failure				
							Written off				
J-359	2426	4/6/40		3.03	11071	7/10/40	Acquisition				
						4/15/42	Taxi maneuver	Taxied into marking borad	Thun	Brügger W	8
					11106	11/2/44	Collision	Wing grazed J-311 in air combat practice	Pilatus Stans		15
				230.43	10764	9/23/47	Normalization	Landing gear released only 45°, retractable cylinder eroded, contaminated hydraulic oil	Buochs	Christeler P	
			8/14/48	229.39	10764		Belly landing				
							Written off	*No longer flown since accident on 9/23/47*			
J-360	2427	12/20/39		3.23	10957	3/18/41	Acquisition				
				87.59	*10691*	3/3/45	Engine change	Engine failure, supercharger defect	F+W Emmen		
			12/28/49		11055		Normalization				
							Written off				
J-361	2428	4/9/40		3.48	11034	4/16/42	Acquisition				
						10/24/42	Takeoff maneuver	Engine failure, front shaft drive broken	Biel-Bözingen	Kilchenmann P	15
						8/22/44	Belly landing	Propeller touched ground	Payerne		9
					10948	10/18/44	Emergency landing	Landing gear defect	Pilatus Stans		
						5/14/45	Normalization	Supercharger defect	Payerne	Burlet G	8
			12/28/49			3/2/49	Propeller shot	Munitions defect			
							Written off				
J-362	2429	12/20/39		3.41	10921	8/29/40	Acquisition				
						10/18/44	Emergency landing	Engine failure	Emmen	Wannenmacher E	21
					10947	12/16/44	Takeoff maneuver	Engine failure, injection pump drive broken		Brenzikofer F	9
			8/14/48	337.09	11045		Normalization		F+W Emmen		
							Written off				
J-363	2430	12/18/39		2.42	11064	4/2/43	Acquisition				
							Takeoff failure	Overturned, heavy damage, in repair until 9/18/43	Belp	Honegger ?	
						3/23/45	Normalization		Pilatus Stans		
			8/14/48	327.08	10762	6/19/45	Emergency landing	Engine failure, oil pump defect	Emmen	Schefer E	21
								Written off			

Aircraft No.	Factory No.	Purchase	Written off	Aircraft Hours	Engine No.	Incident Date	Incident	Comment	Location	Pilot	Fl Kp/St
J-364	2431	4/14/40		3.35	11006	6/8/40 9/21/40 3/29/45	Acquisition Air combat Taxi maneuver Normalization Written off	Emergency landing due to combat damage Taxied into location marking	Olten Payerne Pilatus Stans	Borner A Wachter A	21 6
J-365	2432	4/5/40	8/10/42	2.14 89.30	11058 11058	12/26/41	Acquisition Belly landing Crash	Tire blow-out at takeoff Collision with J-366, parachute jump	Thun Gerzensee	Eggenberger M Rosenmund P	9 6
J-366	2439	12/20/39	8/10/42	2.37 93.30	11040 11040		Acquisition Crash	Collision with J-365, parachute jump	Gerzensee	Brocard A	6
J-367	2440	12/15/39		2.53	10698	2/21/41	Acquisition	Nose dive until max permissible speed, canopy damaged		Eggenberger M	9
			8/14/48	369.32	10694 10772 10917 10769	11/17/43 3/7/44 10/24/44 4/20/45	Engine change Engine change Landing failure Landing failure Normalization Written off	Engine failure, shaft eroded, Swerved, landing gear torn left Swerved, 50,000 francs damage	Locarno Interlaken Meiringen Pilatus Stans	Streiff V Dosch J Bueche J	21 8 6
J-368	2441	12/16/39	6/24/42	3.47 111.28 114.38	10967 10986 10986	1/16/42	Acquisition Engine change	Collision in mountain range	Buochs Pilatus	Wild J (t)	Fl Abt 5
J-369	2442	4/25/40	8/14/48	2.56 314.21	10975 11071	2/6/45	Acquisition Normalization Written off		F+W Emmen		
J-370	2351	12/16/39	8/14/48	3.09 276.48	10956 10921 10921	9/29/44	Acquisition Normalization Written off		Pilatus Stans		
J-371	2352	4/20/40	11/20/46	296.10	10970 11000	5/12/44 2/22/45	Acquisition Landing failure Normalization Written off	Swerved, left wing grazed ground	Payerne F+W Emmen	Dosch J	8
J-372	2353	4/10/40	11/20/46	2.13 316.33	10915 10907	4/20/43 6/10/43 12/11/44	Acquisition Emergency landing Landing failure Normalization	Engine failure, supercharger defect Right wheel blocked while braking, left wheel spokes broke Written off	Belp Thun F+W Emmen	Feldmann F Widmer E	8 9

Aircraft No.	Factory No.	Purchase	Written off	Aircraft Hours	Engine No.	Incident Date	Incident	Comment	Location	Pilot	Fl Kp/St
J-373	2354	4/15/40		2.43	10971	10/28/43	Acquisition		Buochs		
				147.44	*10983*	1/25/45	Parts revision		Pilatus Stans		
						2/28/46	Normalization	Cheval de bois	Sion	Häberlin HP	9
			8/14/48	338.59	10772	10/3/46	Landing failure Propeller shots Written off	Wrong munitions	Locarno	Dannecker F	8
J-374	2355	4/23/40		2.10	10901	9/21/40	Acquisition	Landing gear buckled on ground, hydraulic failure	Payerne		Fl RS
F+W Emmen					10903	10/13/44	Normalization				
			12/28/49		10991	*9/24/45*	Emergency landing Written off	Engine failure, injection pump front shaft drive broken	Seedorf (FR)	Rufer A	9
J-375	2356	*12/2/39*		2.35	10910	3/8/40	Acquisition Landing failure	Other date 12/15/39 Veered off on concrete runway, severely damaged	Geneva	Brenneisen M	8
					11064	2/10/41	Emergency landing	Engine failure, coolant loss	Dübendorf	Mathez M	6
						1/7/45	Normalization		Pilatus Stans		
			8/14/48	356.06	10971	4/25/45	Emergency landing Written off	Engine failure, pressure regulator defect	Interlaken	Meyner H	21
J-376	2357	4/9/40		2.33	11106	7/23/41	Acquisition Taxiing accident	Nose-over, 2800 francs damage	Buochs	Mühlemann E	21
						4/20/42	Engine change	Engine failure, ball bearing defect	Emmen	Aschwanden E	Det 53
				278.35	10975	7/21/42	Emergency landing	Lack of fuel	F+W Emmen		15
						2/27/45	Normalization				
			8/14/48	308.43	10975	11/2/46	Written off	Canopy lost during flight, 2000 francs damage	Locarno	Peyer F	UeG III
J-377	2358	4/27/40			11054	9/13/40	Acquisition		Dübendorf		
					10700	9/28/43	Emergency landing	Elevator blocked, tool – hammer caught	Dübendorf	Kisling R	21
					10772	11/2/44	Engine change Normalization		Pilatus Stans		Det 30
			12/2/49			6/16/48	Emergency landing Written off	Engine failure, ignition coil burnt	Locarno	Burlet G	8
J-378	2359	4/20/40			11051		Acquisition				
			9/5/44	205.25	10904	6/12/44	Takeoff failure Air combat	Nose-over Shooting down by USAAF	Kägiswil Neuaffoltern	Ernst E Treu P (t)	8 7
J-379	2360	4/12/40		2.54	11045	6/20/41	Acquisition Emergency landing	Fuel pressure drop, fuel pump eroded Fuselage rammed	Buochs *Zweisimmen*	Reber A Burlet G	15 6
					11051	*12/17/40* 12/12/44	Normalization				
Pilatus Stans			8/14/48	290.35	10993		Written off				

248

Aircraft No.	Factory No.	Purchase	Written off	Aircraft Hours	Engine No.	Incident Date	Incident	Comment	Location	Pilot	Fl Kp/St
J-380	2361	4/26/40		2.39	10996	4/23/41 3/20/43 7/23/43 11/18/44	Acquisition Emergency landing Emergency landing Emergency landing Normalization	Lack of fuel Engine failure, supercharger defect Drop in oil pressure	Payerne Thun Belp	Ahl A Mathez M Künzler F	21 6 7
F+W Emmen			6/27/46	352.28	10691 21539		Emergency landing	Propeller adjustment defect, aircraft a total loss	Avenches	Egli F	15
J-381	2373	4/25/40	4/26/48	227.05	11080 11080 10907	9/9/41	Acquisition Landing failure Normalization Landing gear defect	Left landing gear dented Overturned after landing with half-lowered landing gear	Olten F+W Emmen Buochs	Frei W Schenk P	8
J-382	2374	4/27/40	12/2/49		11064 11035	10/26/44	Acquisition Normalization Written off		F+W Emmen		
J-383	2375	4/20/40	12/18/49	2.28	11035 10752	2/14/41	Acquisition Landing failure Normalization Written off	Tire blow-out, swerved, Overturned, fuselage 30% damage	Buochs Pilatus Stans	Köpfli M	21
J-384	2376	4/16/40	8/14/48	2.48 45.05 278.57	11081 11046	4/4/41 2/11/44 1/10/45	Acquisition Taxiing accident Normalization Written off	Wing deformation determined, wing change Nose-over, 4000 francs damage	Payerne F+W Emmen	Brenzikofer F	9
J-385	2377	4/16/40	8/14/48	2.48 358.34	11049 10764	9/1/42 2/7/45	Acquisition Emergency landing Normalization Written off	Fuel tank leaky	Lausanne F+W Emmen	Künzler F	7
J-386	2378	4/15/40	8/14/48	2.55 340.41	11037 10695 10695	3/28/45	Acquisition Normalization		Pilatus Stans		
J-387	2379	4/5/40	12/28/49	2.07	11086 10915 10915	3/28/44 12/27/44	Acquisition Engine change Normalization Written off		Interlaken Pilatus Stans		
J-388	2380 15	4/6/40		2.13	11095 10970	4/22/41 4/22/42	Acquisition Engine Engine change	Valve spring break Swarf or cloths in the oil filter	Interlaken		

249

Aircraft No.	Factory No.	Purchase	Written off	Aircraft Hours	Engine No.	Incident Date	Incident	Comment	Location	Pilot	Fl Kp/St
J-389	2381	4/12/40	12/28/49		10946	12/21/44	Normalization Written off		Pilatus Stans		
J-390	2392	4/20/40	8/14/48	2.31 280.52	10984 11049 10773	9/26/44	Acquisition Normalization Written off		Pilatus Stans		
J-391	(2301)	4/28/44	12/28/49	3.09	10947 11054	2/5/45	Acquisition Normalization Written off	Test MG cooling (date unknown) Assembly through DMP Buochs, Me 109 D fuselage (J-306) and replacements parts Me 109 E	Pilatus Stans		
				62.54	10762	7/12/44	Normalization	Collision with drag chute cable	Payerne	Fischer A	8
			3/28/47	147.37	11071	5/15/45	Crash landing	Aircraft oversteered, swerved, 182,000 francs damage	Payerne	Witmer K	9
J-392		7/18/45				4/24/46		Assembly from spare parts, Doflug Altenrhein After takeoff canopy lost	Thun	Scherer R	8
						1/20/49		Tire blow-out at takeoff Written off	Buochs	Brenneisen M	8
J-393		9/10/45	12/28/49			6/18/48	Emergency landing	Assembly from spare parts through DA* Propeller adjustment defect, engine oversped	Samedan	Brenneisen M	8
J-394		12/7/45	12/28/49				Written off	Assembly from spare parts through DA*			
J-395		12/7/45	5/22/47	61.01	10691	8/13/46	Takeoff failure	Crosswing, wing grazed ground Collision with fir tree due to poor visibility	Interlaken Montfaucon	Schwarzenbach R Bueche JP (t)	8 8
						1/15/46		Assembly from spare parts through DA* Collision with chute cable	Locarno	de Brémont C	8
						11/5/47	Emergency landing	Drop in oil pressure, line broke on manometer Written off	Payerne	Dannecker F	
J-396		12/7/45	12/28/49			10/3/46	Emergency landing	Engine defect	Locarno	Dannecker F	8
						4/3/47	Emergency landing	Propeller at high gradient due to	Locarno	Loup G	A Fl Pk oil loss
J-397		12/21/45	12/28/49			5/3/49	Aborted flight Written off	Engine failure, break in cylinder	Thun	Ernst E	8
J-398		1/10/46 12/28/49	12/28/49				Written off	Assembly from spare parts through DA* Assembly from spare parts through DA*			
J-399		3/19/46	12/28/49			1/21/48	Landing	Assembly from spare parts through DA* Cheval de bois, left tire blow-out Written off	Payerne	Scherer R	8
J-701	163112	5/20/44			00209676 109985	5/25/46	Crash landing	Landing gear not dropped Written off	Altenrhein	Kind C	7
J-702	163320	5/20/44	9/8/47		01104624						
J-703	163243	5/20/44	9/8/47		01104652		Written off				

250

Aircraft No.	Factory No.	Purchase	Written off	Aircraft Hours	Engine No.	Incident Date	Incident	Comment	Location	Pilot	Fl Kp/St
J-704	163245	5/20/44			0104217	8/26/44	Emergency landing	Engine defect, valve plate break	Dübendorf	Siegfried H	7
						1/3/45		Aircraft swerved guidless	Interlaken	Läderach W	DMP
						10/22/45		Canopy lost during flight	Basel	Widmer E	7
						11/13/46		Engine defect, smoke in cabin	Interlaken		A Fl Pk
J-705	163248	5/20/44	2/7/47	57.38	0020 790		Emergency landing		Sion	Hofer R	7
					00202165	11/22/45	Written off	Expansion in fuel tank, supply interrupted			
J-706	163251	5/20/44	9/8/47		00203790	8/4/44	Emergency landing	Engine supercharger defect	Interlaken	Feldmann F	7
						6/14/46	Takeoff failure	Aircraft swerved, left shock strut constricted	Buochs	Mirault G	UeG
			9/8/47				Written off				
J-707	163084	5/23/44			01102872	5/30/45	Emergency landing	Engine supercharger defect	Interlaken	Mathez M	6
						6/26/46	Landing failure	Touched down too late, grazed shed with wing	Emmen	Rageth E	7
						10/25/45		Fuel tank casing torn out	Lodrino	Schoch A	7
			5/28/48			11/20/46	Emergency landing Written off	Engine supercharger defect	Payerne	Hofer W	7
J-708	163806	5/23/44	9/8/47		01102929		Written off				
J-709	163808	5/23/44	9/8/47		00203253	5/27/46	Taxiing accident	Taxied into runway marking Written off	Dübendorf	Meyner H	A Fl Pk
J-710	163814			106.11	00703871 703871	4/13/46	Emergency landing	Total loss, propeller adjustment pawl broken	Buochs	Wiesendanger F	7
			10/2/46				Written off				
J-711	163815	5/23/44	9/8/47		01103896	1/9/45	Engine defect	Locking pin on spark plug nipple fell out Written off			
J-712	163816	5/23/44	9/8/47		00703975	8/19/44	Crash landing Written off	Right landing gear buckled			
J-713	162764	12/45	5/29/46	18.07	01102434 01104652		Crash	ex Me 109 G-6 / J Bo Rei (RU+OZ) Aircraft and pilot found on 9/4/53	Interlaken	Treu P	7
J-714	482818	1946			01101442		Written off	ex Me 109 G-14 (Factory No. 462818)	Pizzo di Rodi	Zweiacker H (t)	7
J-715	7605	(1946)	5/28/48					Me 109 F-4/Z The aircraft no. was assigned in June 1946. The aircraft never came to the troop.			
J-716	7197	(1946)	5/28/48					Me 109 F-4 The aircraft no. was assigned in June 1946. The aircraft never came to the troop.			

Comments:
The purchase dates normally mean the transfer of the aircraft to the troop (for example KTA to the Armeeflugpark). This was mostly synonymous to the overflight from Germany.
In flight statistics at the end of 1938 two Me 109 D are found. However, the purchase dates indicate a single aircraft.
The Me 109 F was not flown by the troop (Armeeflugpark). There are indications that an aircraft was flown at the beginning of the year 1943 by F+W Emmen (KTA ?).
*DA = Doflug Altenrhein
The dates come from sources such as "Flugzeug-Chronik der Schweizer Militäraviatik," "Flugstatistik," personal documents from J. Urech.
Discrepancies are possible. Dates in *cursive* are not confirmed.

Abbreviations

Abbreviation	Term	Comment
ANR	Aeronautica Nationale Repubblicana	Italy's air force on the German side as of September 1943
A.W.Z.	Auswertezentrale	
AFLF	Abteilung Flieger- und Fliegerabwehr	
Cp av	Compagnie d'aviation	See under *Fl Kp*
DMP	Direktion der Militärflugplätze	Later BABLW (*Bundesamt für Betriebe der Luftwaffe*) or RUAG Aerospace
Doflug	AG für Dornier Flugzeuge, Altenrhein	as of 1948 FFA (*Flug- und Fahrzeugwerke Altenrhein*)
EKW	Eidg. Konstruktionswerkstatt Thun	Later K=W Thun, today RUAG
EMD	Eidg. Militärdepartment	Today VBS (*Verteidigung, Bewölkerungsschutz, Sport*)
EW/EWZ	Escher-Wyss Zürich	Today Sulzer-Escher Wyss AG
F+W	Eidg. Flugzeugwerk Emmen	Today RUAG Aerospace
FG	Fighter Group (USAAF)	*Fliegereinheit* with 3 to 4 Fighter Squadron
Fl Kp	Flieger Kompanie	Until 1945 term for *Fliegereinheit*
Fl St	Flieger Staffel	As of 1945 term for *Fliegereinheit*
Fl.B.M.D.	Flieger Beobachtungs- und Meldedienst	
Flab	Fliegerabwehr	In German Flak
FPD	Eidg. Flugplatzdirektion	as of 1928 *Kommando Fliegerwaffenplatz*, 1938 DMP
FS	Fighter Squadron (USAAF)	*Staffel* with 12 to 16 aircraft
JG	Jagdgeschwader (Germany)	3 to 4 *Gruppen* à circa 27 aircraft
K+W	Eidg. Konstruktionswerkstatt Thun	See EKW
KTA	Kriegstechnische Abteilung	Today *Armasuisse*
Mtt	Fa. Messerschmitt	
PS	Pferde-Stärke	Today kW (1000 PS = 735.5 kW)
RAF	Royal Air Force	Great Britain's air force
Reduit	Stellung der Armee im Alpenraum	
RLM	Reichs-Luftfahrts-Ministerium	Aeronautical authority of Germany
SLM	Schweiz. Lokomotiv & Maschinenfabrik Winterthur	
UeG	Überwachungsgeschwader	*Geschwader* with professional pilots 1941-2005
USAAF	US Army Airforce	USAF since 1947

Index

(Cursive = photograph)

1. People

Aschwanden E, Pilot *Fl Kp* 15: 99, *100*, 102, 103, 105, 106, 234, 139, 244, 248
Bandi H, *Kdt FF* Trp until 1943: 29, 31, 32, 34, 35, 36, 40, 54, 72, 84, 85, 86, 113
Becker Reinhold, *Waffenkonstrukteur*: 45
Benoit JP, Pilot Cp av 6: 68, 108, 111, *122*, 240, 241
Biozzero E, Pilot *Fl Kp* 9: 234
Birkigt Marc, Hispano-Suiza: 11
Borer A, Pilot *Fl Kp* 14: 105, 106
Borner A, Pilot *Fl Kp* 21: 108, 111, *124*, 244-146
Brenzikofer F, Pilot, *Fl Kp* 9: 66, *126*, 241, 246, 249
Dewoitine Emile: 12, 26, *27*
Duttweiler G, *Nationalrat*: 30, 31
Eggen von, Rittmeister, Diplomat: 150, 151
Egli F, Pilot *Fl Kp* 15: 105, 106, 108, 110, *141*, 237, 242, 248
Fischer A, Pilot *Fl Kp* 8: 115, 234, 239, 242, 249
Frei Wilhelm, Pilot *Armeeflugpark*: 91, *92*, 238
Furrer Adolf, *Waffenkonstrukteur*: 39
Göring Hermann: Quote, 12, 15, 34, 76, 86, 104, 107, 151
Greim von, Ritter, dt. General: 104, 108
Gripp, dt. *Militärattaché*: 104, 108
Guisan Henry, *Oberbefehlshaber* of the army: Title, 84, 90, 112, *118*, 161
Gürtler E, Beobacher *Fl Kp* 10: 108
Hadorn E, Pilot Cp av 6: *99*, 108, 111, *128*, 244
Heiniger E, Pilot *Fl Kp* 13: 106
Heiniger R, Pilot *Fl Kp* 7: 132, *132*, 133-136, 174, *194*, 240
Heinkel Ernst: 12, 15, 17, 28, 30, 56
Hitz P, Pilot *Fl Kp* 9: 79, 106, 234
Hofer HR, Tech *Of Fl Kp* 7: *194*
Hofer R, Pilot *Fl Kp* 7: 164, *166*, 171, *205*, 237, 251
Högger K, Pilot *Armeeflugpark*: 22, 85, 148
Homberger R, Pilot *Fl Kp* 15: 87, 99, 105, 108, *109*, 110, 111, *113*, *201*, 237, 241
Hörning W, Pilot *Fl Kp* 21: 95, *96*, 97, 108, 111, *124*, 234, 239
Johnen W, dt. Pilot NJG 5: 150
Junkers Hugo: 56
Kind C, Pilot *Fl Kp* 7: 168, *168*, 250
Kisling R, Pilot *Fl Kp* 21: 96, 237, 248
Köpfli M, *Fl Kp* 21: *90*, *125*, 238, 241-243, 248
Koschel H, *Kdt Fl Rgt* 2: 137
Kuhn H, Pilot *Fl Kp* 15: 99, 105, 106, 108, *109*, 110
Künzler F, Pilot *Fl Kp* 7: 132-134, 241, 243, 248, 249
Läderach W, Pilot *Fl Kp* 7: 57, 58, 70, 115, *122*, 132, 134, 135, 164, *206*, 234, 237, 238, 250
Liardon F, Pilot Cp av 6: 108, 111, 115, *128*, 234, 241-243
Lindecker W, Pilot *Fl Kp* 15: 41, 59, 99, 100, *100*, 102, 103, 108, 110, 115, 234
Messerschmitt Willy: 14, *14*, 15, *16*, *18*
Meuli R, Pilot *Fl Kp* 10: 108
Meyner H, Pilot *Fl Kp* 21: 64, *124*
Mooser, Pilot KTA: 57
Morier H, Pilot Cp av 2: 104
Mühlemann E, Pilot *Fl Kp* 21: 108, 240, 248
Mutke G, dt. Pilot: 150

Nipkow G, Pilot *Fl Kp* 9: 106, 239
Preiss F.A.: 31
Rickenbacher R, Pilot *Fl Kp* 15: 106, *106*, 107, *113*, 238
Rihner F, *Kdt FF Trp* as of 1944: 66, 79, 86, 137
Ris W, Pilot *Fl Kp* 14: 106
Roubaty J, Pilot Cp Av 6: 98, 125, 234, 240
Rufer A, Pilot *Fl Kp* 15/*Fl Kp* 9: 65, *100*, 104-106, 115, *126*, 186, 237, *237*, 241, 247
Scharf H, dt. Pilot: 144, *145*
Scheitlin R, Pilot *Fl Kp* 21: 108, 111, *124*
Schenk P, Pilot Cp av 6: 99, *128*, 240, 248
Schneider Franz: 11
Schoch A, Pilot *Fl Kp* 7: 132-134, 251
Siegfried H, Pilot *Fl Kp* 7: 134, 194, 241, 250
Streiff V, Pilot *Fl Kp* 21: 64, 65, 96, 97, *98*, 108, 111, 115, *124, 201, 243, 238, 244, 246*
Suter R, Pilot *Fl Kp* 15: 61, *100*, 106, 107, 234, 238, 243
Thurnheer H, Pilot Cp av 6: 57, 95, *96*, 99, 108, 111, *128*, 237
Treu P, Pilot *Fl Kp* 7: 78, 132-136, 161, 162, 248, 251
Udet Ernst: 12, 13, 16, 17, 19, *19*, 20, *20*, 23, 24, *24*, 25, 34
Villing M, dt. Pilot: 144, *145*
Wachter A, Pilot Cp av 6: 98, 244, 246
Wannenmacher E, Pilot *Fl Kp* 21: 64, *64, 124*, 248, 240, 242, 246
Willi G, Pilot *Fl Kp* 21: 108, 243
Widmer E, Pilot *Armeeflugpark*: 171, *206*, 239, 243, 244, 247, 250
Witmer K, Pilot *Fl Kp* 9: 249
Wiesendanger F, Pilot *Fl Kp* 7: 134, 164, *167, 194*, 205, 234, 243, 251
Wittwer M, Pilot *Fl Kp* 13: 106
Zweiacker H, Pilot *Fl Kp* 7: 170, 251

2. Units

Cp av/ESC av 6: 46, *52, 59*, 54, *68*, 70, 90, *94*, 95, 98, *99*, 111, 115, 117, *122, 125, 128, 180, 183*
Fl Kp/Fl St 7: 54, 75, 91, *92, 116, 118-122, 130*, 132, 135, 142, 161, 162, 164, *164, 169, 170, 171, 174*, 184, *189*, 194
Fl Kp/Fl St 8: 54, *88*, 115, *117*, 119, *121-123, 137, 139, 140*, 142, *185, 199*
Fl Kp/Fl St 9: 54, 55, 55, 66, *104*, 106, 115, *118*, 119, 123, *126, 128, 138, 141, 186, 193, 196*
Fl Kp/Fl St 15: 41, *56*, 65, 66, *66*, 90, *92*, 95, 99, *100*, 102, 105, *105*, 106, 108, *108*, 110, *110*, 111, 112, *113*, 115, *124, 127, 138, 141, 187, 189*
Fl Kp/Fl St 21: *44*, 64, *64*, 75, *89*, 90, *91*, 93, *95*, 96, *96*, 97, *98*, 108, 111, *111*, 115, *125, 179, 180, 198*, 142
UeG: 79, 80, 116, 117, 132, 137, 161, *169*
Fl.B.M.D.: 91, 94, 96, 135
Groupe de Chasse II/7: 98, 101, 102
JG 2: 144, *144, 146*
JG 3: *157*, 158
JG 5: 144, 145, 149
JG 26: 144, *144, 146*
JG 77: 156, *156*
KG 27: 96, 97
KG 51: 95, 96
KG 53: 98, *98*, 99, 101
KG 55: 101, 102, 104, 107

253

ZG 1: 57, 104, 105, 106, 108, 110-112
USAAF 339th FG: 132, 134-137

3. Locations

Affeltrangen: *157*, 158
Altenrhein Airfield: 43, 61, *67*, *82*, 83, *86*, 96, *140*, *150*, 164, *168*, *170*, *178*
Avenches, Airfield: 46, *46*, *75*, 95, 106, *116*, *118*, *122-125*, *141*, *183*, *184-186*, *196*, *Appendix 19*
Bellinzona: 20, *21-23*
Bern-Belp: 86, 95, 144, *144*, 145
Bern-Beundenfeld: *156*
Biel-Bözingen: 57, 95, *104*, 106, *108*, 110, 111
Boécourt: *107*, 195
Brugg: 95
Buochs Airfield: 24, 54, 55, *55*, *67*, 68, *70*, *71*, 81, *81*, 82, 102, *103*, 110, *125*, *129*, *142*, 150, 162, 164, 165, *166*, *167*, *169*, *176*, *178*, *198*
Bütschwil: 95
Dent de Vaulion: 104
Doubs: 99, 100, 104, 105
Dübendorf Airfield: 5, 16-21, 23, *28*, 29, 32, 41, *42*, *56*, *57*, *58*, *59*, *59*, *60*, 61, *62*, *63*, 64, 66, 77, 81, *87*, 89, 90, 91, *91*, *93*, 95-97, 108, *130*, *131*, 132, 133, 134, *134*, 135, 136, *139*, 140, *140*, 150, 151, *151*, 153, *153*, 156, 161, 176, *176*, *177*, *178*, *180*
Emmen, Airfield: *44*, 95, *117*, *125*, *131*, 145, 146, *166*, 170, *188*
Forel, flyer shooting range: 65, *142*, 161, 174, *189*, *195*
Friedrichshafen, Airfield: 37, *37*, *56*, 61
Grande Combe des Bois/Le Russey: 106
Genf, Airfield: 91, *92*, 95, *102*
Interlaken, Airfield: *47*, *126*, 132, *138*, *154*, 161, *162*, 164, 165, *171*, 174, *180*, *Appendix 20*
Kägiswil: 66
Kemleten: 96, *97*
La Chaux-de-Fonds: 105-107
Lausanne-Blécherette, Airfield: 84, 85, 104, 106
Les Rangier: 100
Lignières: 98, *98*
Mollis, Airfield: 64, 95
Oberkirch/Nunningen: *111*, 112
Oltingue: 99
Olten, Airfield: *93*, 95, 96, 99, *99*, *100*, 105, 106, 108, 110, 111, *113*, *124*, *189*
Payerne, Airfield: 41, *47*, 65, *78*, *80*, *82*, 90, 95, *138*, 161, 171
Pontarlier: 104
Porrentury/Pruntrut: 137
Saingnelégier: 99, 106, 107, 111
Samedan: 149
Sion: 164, *166*
St. Ursanne: 105, 110
Thun, Airfield: 20, *24*, *26-28*, 30, *52*, *53*, *68*, *73*, 90, 95, 98, 99, 111, *125-127*, *161*, 162, *180*, *182*, *198*
Triengen: 111, *111*
Ueberstrass: 111
Unterbach, Airfield: 66, *137*, *Appendix 22*
Ursins: *102*, 102
Zweisimmen, Airfield: *128*, *Appendix 21*

4. Aircraft

Arado Ar 96: 149, 193
B-17G "Blues in the Night": 132-134, *132*

B-24H "Lonesome Polecat": 132, 134
Bell P-39 Airacobra: 83
BFW M-18: 14, *33*
Breda Ba.65: 32
Breda Ba.88: 32, 33
C 35: 21,22, *22*, 23, 28, 30, 32, 33, 36, 51, 80, 89, 91, *93*, 94, 104, *107*, 108, 112
C 36/C 3603: 28, 53, 76, 77, 80, 125, 137, 145, 171, *171*, 194
Caproni Ca 310/312: 83
Curtiss Hawk H-75A-1: 31
Curtiss Hawk II: 24
Curtiss Hawk 81-A: 83
Curtiss P-36: 31, 33
CV: 26, 30, 36, 41, 51, 81, 89
D 3800/01: 51-54, 72, 73, 76, 77, 78, 80-84, *84*, *95*, *98*, *101*, *106*, *122*, 130, 142, 145, 171, 194
D 3802/02A/03: 86, *87*
Dewoitine D.19: 39
Dewoitine D.27: 23, 26, *27*, 30, 36, 41, 51, 59, 89, *89*, 95
Dewoitine D.500/501: 11
Dewoitine D.510: 20, 21, *21*, 22, *23*
Dewoitine D.513: 28
Dewoitine D.520: 98, 101, 102
Dewoitine D.9: 39
Dornier Do 17: 17, 21, *21*, 105
Douglas 8A: 33
Fairey Fox: 21-23
Fairey P.4/34: 33
Fiat CR.32: 23
Fiat G.50: 11, 32, 33, 145
Häfeli DH 5: 39, 51
Hawker Fury: 23
Hawker Henley: 33
Hawker Hurricane I: 11, 33, 72
Hawker Sea Fury: 87, *87*, 174
Hawker Tornado: 83
Heinkel He 111: 45, 91, 95, 96, 97, *97*, 98, *98*, 100, 101, *101*, 102, *103*, 104-107, *Appendix 17 + 18*
Heinkel He 112: 11-13, 17, 22, *23*, 28, 29, *29*, 30, 32
Junkers Ju 88: 95
Koolhoven FK 55: 28, 29
Loire 250: 28
Macchi C 200: 83
Macchi C 202: 84, 85
Macchi C 205: 84, *85*, *140*
Me 109 V13/D-IPKY: 17, 21, 22
Me 109 V14/D-ISLU: 16, 17, 18, *18*, 19, 20, *20*, 24, *25*
Me 109 V7/D-IJHA: 17, *18*, 19, *19*
Me 109 V9/D-IPLU: 17, *17*
Me 109B-1 Factory No. 1062: 17, *17*, 18
Me 110:30, 32, 35, 57, 85, 104-108, 109-112, *111*, *112*, 137, *140*, 150, *150*, 151
Messerschmitt Me 262: *140*
Morane-Saulnier MS 227: 11
Morane-Saulnier MS 405: 11, 19, 21, 28, 30-32
Morane-Saulnier MS 406: 31, 33, 72, 83, *90*, 95
Morane-Saulnier MS 540: 86
Mosquito PR IV: 145
Nardi 305/315: 32, 33, 35
Nieuport Ni 160: 28
North American NA-73: 83
P-51B Mustang: 132, *133*, 134
P-51D Mustang: 84, 87, *88*, 142, 171, 188

254

Pilatus P-2: 193, *193*
Potez 25: 51, *63*
Potez 63: 30, 32, 91, *94*
Reggiane Re 2000: 83
Republic P-43 Lancer: 83
Spitfire: 33, 83, 149
Vought V-156: 33
Vought XF4U-1 Corsair: 83
Wasserflugzeuge: 81

5. Engines and Propellers

Chauvière 351 Propeller: 72
Daimler E4uF: 11
DB 600: 17, 24, *28*, 30
DB 601:13, 17, 18, *24*, 32, 38, 39, *71, 72, 73*, 74, *75, 84, 85, 144, 148, 194*
DB 605: 13, 85, 155, 159, *163*, 171, *172*, 194
EW V6 Propeller: 72, *82, 137, 198, Appendix 16*
HS 12Xcrs: 11, 29
HS 12Ycrs: 29, 30
HS 51: 73, 83
HS 76: 11
Jumo 210: 12, 13, 17, 29, 30, 32, 38, 39
Propeller shots: 80
Saurer YS 2/YS-3: 87
VDM Propeller: 72, *124, 174*

6. Weapons and Weapons Assemblies

Wild R-V1: *40*, 41
FF-K: 39, 45, *45, 46, 97, 103, 182, Appendix 7-9*
CC-Gear: 39
Revi 3c:41, 43, *49*, 61, *Appendix 12*
Revi 12/b: *154*
FM-K 38: 39, 45, 104
MG 131: 13, 154, 194
MG 151/20: 13, *147, 148*, 154, 194
MG 17: 13, 18, 38-41, 43, 44, *105, 147*
MG 29: 39-41, *42*, 43, *43, 44, 49*, 61, 80, 154, 193, 194, *Appendix 3-6*
MG C/30L: 39

MG-FF: 13, 39, 41, 45
Bombs/Bomb Assemblies: 28, 30, *73*, 77, *77*, 78, 80, 146, 156, 193, *Appendix 14*
Additional tank: 156

7. Radio Equipment

Radio equipment FG IX: 52, *53*, 54, 55
Radio equipment FG VII: 51, *51*, 52, 53, 54
SIF 450: see FG IX
T.S.F: see FG IX
Radio equipment FuG 16 Z: 154, *161*
Radio equipment FuG 25a:150, *153*, 154, 155, *161*
Radio equipment General: 30, 36, 51, 53, 135, 154

8. General

Arado: 12, 56
Fuel, K-Kraftstoff: 73, 74
Bührle Fa: 83, 66
Dornier Fa: 17, 43, 61, 83, *101, 67, 82*
F+W Emmen: 15, *50*, 54, 77, 78, 79, 82, *131*, 145, *148*, 150, 161, 164, 181
Fiat Turin:32, 171
Fliegermuseum Dübendorf: 193, 194
Concentration camp, Concentration camp prisoners: 12, 14, 51, 152
Aviation meeting, Dübendorf 1937: 9, 16, 29
France (aircraft procurement): 30-34, 36, 53, 54, 83, 87
Hispano-Suiza: 11, 29, 33, 145
Italy (aircraft procurement): 31, 32, 38, 76, 83, 85
K+W Thun: *27, 28, 29*, 41, 53-55, 72, *84, 90*
Catapult device: 81, *81*
Maschinenfabrik Oerlikon: 45, 46, 83
Night flight equipment: 30, 79, *79*, 80
Armor protection: 87, 78, 79
Pilatus Fa: 83
Oxygen system: 40, 76
Signal rockets: 53, *73, 79, Appendix 13*
Control stick KG11: *48, Appendix 10-11*
Wing deformations: 36, 65, *65*, 66, *66*, 78, 161, *161*
W+F, Bern: 39, 45

255

Bibliography

Author	Title	Publisher	Year
Aders G	Geschichte der Deutschen Nachtjagd	Motorbuch Verlag, Stuttgart	1977
Bürli W	Flugzeugbewaffnung	Stocker-Schmid, Dietikon	1994
Cross/Scarborough	Messerschmitt Bf 109 B-E: Their History and how to model them	PSL/AIRFIX, London	1972
Cuney/Danel	Dewoitine D.520	DOCAVIA, Paris	1974
Cuney/Danel	Les Avions Dewoitine	DOCAVIA, Paris	1982
Dierich W	Kampfgeschwader 51	Motorbuch Verlag, Stuttgart	1991
Dierich W	Kampfgeschwader 55	Motorbuch Verlag, Stuttgart	1994
Fl Staffel 15	Die Geschichte einer Fliegerstaffel	Fl Staffel 15	1967
Fl Staffel 21	Fliegerstaffel 21 1946-1993	Fl Staffel 21	1993
Fl Staffel 7	7-ner Post	Fl Staffel 7	1984
Fleig K	Der Flugzeugmaler	De Grueter & Co, Berlin	1944
Freeman R	The Mighty Eight	Arms & Armour Press, UK	1990
Hoffmann H	Me 109 – der siegreiche deutsche Jäger	Hoffmann, München	1941
Kiehl H	Kampfgeschwader 53	Motorbuch Verlag, Stuttgart	1996
Krause E	Der Metallflugzeugbauer	Teubner, Leipzig und Berlin	1943
Mettler E	Die schweizerische Flugzeugindustrie von den Anfängen bis 1961	Polygraphischer Verlag AG, Zürich	1966
Price A	Luftkampf, Taktik und Technik	Stalling, Oldenburg	1976
Prien/Rodeike	Messerschmitt Me 109 F,G&K Series	Schiffer, Atglen, USA	1993
Rutschmann W	Die Schweizer FF Truppen 1939-45	Ott Verlag, Thun	1989
Schliephake M	Flugzeugbewaffnung	Motorbuch Verlag, Stuttgart	1977
Stapfer/Künzle	Strangers in a strange land USA	Squadron/Signal, Carrolton TX	1992
Tilgenkamp E	Schweizer Luftfahrt Band I-III	AERO-Verlag, Zürich	1941/43
Urech J	Die Flugzeuge der schweizerischen Fliegertruppe seit 1914	Th.Gut, Stäfa	1974
Vann F	Willy Messerschmitt	PSL, Sparkford, UK	1993
Wetter E	Geheimer Nachtjäger in der Schweiz	ASMZ, Frauenfeld	1989
Wetter E	Duell der Flieger und Diplomaten	Huber, Frauenfeld	1987
Wyler E	Chronik der Schweizer Militäraviatik	Huber, Frauenfeld	1990

Report of the head of the *Generalstab* of the army to the *Oberbefehlshaber* of the army concerning active duty 1939–1945
Report of the *Kdt FF Trp* to the *Oberbefehlshaber* of the army concerning active duty 1939–1945
Bundes-Archiv Bern, Documents
Fliegermusuem Dübendorf, Documents
Officer's Budget, various volumes